Indigenous People and Politics

AF763807

Edited by
Franke Wilmer
Montana State University

A Routledge Series

Indigenous People and Politics

Franke Wilmer, *General Editor*

Inventing Indigenous Knowledge
Archaeology, Rural Development, and the Raised Field Rehabilitation Project in Bolivia
Lynn Swartley

The Globalization of Contentious Politics
The Amazonian Indigenous Rights Movement
Pamela L. Martin

Cultural Intermarriage in Southern Appalachia
Cherokee Elements in Four Selected Novels by Lee Smith
Kateřina Prajznerová

Storied Voices in Native American Texts
Harry Robinson, Thomas King, James Welch, and Leslie Marmon Silko
Blanca Schorcht

On the Streets and in the State House

American Indian and Hispanic Women and Environmental Policymaking in New Mexico

Diane-Michele Prindeville

Routledge
Taylor & Francis Group
NEW YORK AND LONDON

Published in 2004 by
Routledge
711 Third Avenue,
New York, NY 10017
www.routledge-ny.com

Published in Great Britain by
Routledge
2 Park Square, Milton Park,
Abingdon, Oxfordshire OX14 4RN
www.routledge.co.uk

Routledge is an imprint of the Taylor & Francis Group

Transferred to Digital Printing 2005

First issued in paperback 2012

Library of Congress Cataloging-in-Publication Data

Prindeville, Diane-Michele.
On the streets and in the state house : American Indian and Hispanic women and environmental policymaking in New Mexico / by Diane-Michele Prindeville.
p. cm. — (Indigenous people and politics)
Includes bibliographical references (p.) and index.
ISBN 0-415-94689-1 (alk. paper)
1. Indian women civic leaders—New Mexico. 2. Hispanic American women civic leaders—New Mexico. 3. Women in politics—New Mexico. 4. Women legislators—New Mexico. 5. Environmental policy—New Mexico. 6. Environmental justice—New Mexico. 7. New Mexico—Environmental conditions. 8. New Mexico—Politics and government. I. Title. II. Series.
E78.N65 P75 2003
333.7'09789—dc21 2003012340

ISBN 13: 978-0-415-65052-6 (pbk)
ISBN 13: 978-0-415-94689-6 (hbk)

In memory of my father, Charles Trego Prindeville, Jr.

This book is dedicated to both of my mothers, and to my son Nicolas-Etienne.

Contents

List of Tables	ix
Acknowledgments	xi
Chapter 1: Introduction	3
Why Study the Politics of American Indian Women and Hispanas?	4
Why Study Environmental Politics in New Mexico?	6
The Purpose and Scope of the Research Project	7
A Brief History of Women in New Mexico	13
Women's Political Participation in America	23
Political Science and the Politics of American Indian and Hispanic Women	26
Chapter 2: Design of the Research Project	39
Project Design and Use of the Grounded Theory Approach	40
Selection of the Project Participants	43
Racial/Ethnic Composition of Project Participants	45
Data Collection: The Use of Personal Interviews	49
Engaging in the Interview Process	51
Gender, Race, Ethnicity, and Cultural Considerations	52
Chapter 3: New Mexico's Environmental Justice Movement	57
The Emergence of the Environmental Justice Movement	59
The Environmental Justice Movement in New Mexico	63
Grassroots Politics: Fertile Ground for Women's Leadership	67

CHAPTER 4: WOMEN'S PATHS TO LEADERSHIP 71

Who are New Mexico's Indigenous and Hispanic Policy Leaders? 72
Leaders' Experiences of Political Socialization 85
Motives for Leaders' Political Participation 93
Leaders' Perceptions of Politics and Politicians 103
Acquiring Power: Paths to Political Leadership 106

CHAPTER 5: THE ROLE OF IDENTITY AND POLITICAL IDEOLOGY 111

Politics and the Intersection of Race/Ethnicity and Gender 112
Women, Politics, and Gender Consciousness 124
Feminism and the New Mexico Leaders 127
Environmentalism and the New Mexico Leaders 141

CHAPTER 6: PRACTICING THE POLITICS OF ENVIRONMENTAL JUSTICE 151

Gender and Public Policy 156
Influencing Public Policy, Practicing Environmental Politics 159
Strategies for Effecting Policy Change 169

CHAPTER 7: CONCLUSIONS 181

Summary of the Research Findings 182
Implications of the Findings 189

APPENDIX A: ROSTER OF PROJECT PARTICIPANTS 193
APPENDIX B: ACTIVIST INTERVIEW GUIDE 195
APPENDIX C: PUBLIC OFFICIAL INTERVIEW GUIDE 199
APPENDIX D: HUMAN SUBJECTS REVIEW CONSENT FORM 203

NOTES 205
REFERENCES 223
INDEX 251

List of Tables

1. Racial/Ethnic Identity of Leaders and Position Held 44
2. Racial/Ethnic Identity of Leaders and Women in New Mexico 46
3. Focus and Scope of Activists' Organizations 47
4. Position Held by Public Officials 48
5. Age of Leaders in Study and Women in New Mexico 72
6. Educational Attainment of Leaders and Women in New Mexico 73
7. Household Income of Leaders 74
8. Household Income of Leaders and New Mexico Public 75
9. Leaders' Family Status 80
10. Native Language of Leaders and New Mexico Public 84
11. Factors Influencing Leaders' Political Socialization 86
12. Length of Leaders' Public Service 92
13. Leaders' Motives for Political Participation 94
14. Leaders' Political Party Affiliation 103
15. Factors Indicating Leaders' Racial/Ethnic Identity 114
16. Factors Indicating Leaders' Gender Identity 119
17. Leaders' Identification with Feminism 131
18. Leaders' Identification with Environmentalism 142
19. Basis of Leaders' Environmental Beliefs 146
20. Leaders' Policy Initiatives 160
21. Leaders' Policy Agendas Promoting Human Rights and Self-Determination 161
22. Leaders' Policy Agendas Promoting Community Planning and Development 164

23. Leaders' Policy Agendas Promoting Conservation of Natural Resources and Environmental Quality 167
24. Types of Strategies Used by Leaders to Influence Public Policymaking 174
25. Classification of Strategies Used by Leaders 175
26. Composition of the New Mexico State Legislature by Sex (1998) 190
27. Sex of Candidates for Top Elected Offices in New Mexico (1998) 190

Acknowledgments

TO ALL OF THE WOMEN WHO PARTICIPATED IN THIS RESEARCH PROJECT; through your work you demonstrate love, courage, strength, and vision. Your efforts to empower others, to protect the environment, and to improve the lives of the people of New Mexico help make this a land of enchantment.

To Christine Marie Sierra, my mentor and dissertation chair, for her continuing support and encouragement over the years.

Thanks to the Office of Graduate Studies and to the Center for Regional Studies at the University of New Mexico, and to the Social Science Research Council and the Inter-University Program for Latino Research for their funding of my project.

On the Streets and in the State House

Chapter 1
Introduction

> If we are to reconstruct theoretical accounts of society by seriously including women, we must begin to situate each individual woman's life story in its specific social and historical setting and show how women's actions and consciousness contribute to the structuring of social institutions. We need to go directly to women to learn about their part in the production and reproduction of society. We cannot have adequate theories of society without them (Anderson et al. 1990, 106).

American Indian and Hispanic women are increasingly filling leadership roles in state, local, and tribal politics (Jacquez-Ortiz 1995; Hardy-Fanta 1993; Jaimes 1992; Ford 1990; Gunn Allen 1986).[1] As elected or appointed officials and grassroots organizers, they shape public policy through the agenda-setting and decision-making processes to enhance the quality of life in communities of color.[2] More importantly, while they work to improve social, economic, and environmental conditions for their communities, American Indian and Hispanic women leaders broaden the scope of political debate, increase citizen involvement, and raise the social consciousness of participants (Bretting and Prindeville 1998; Prindeville and Bretting 1998; Pardo 1995; Di Chiro 1992). As they advocate for the rights of community members, these leaders empower them to become an active and informed citizenry—an essential component of a de facto democracy. In effect, these leaders make politics more inclusive. This is especially the case in New Mexico, where American Indian and Hispanic women are key actors in the environmental justice movement and they exert influence in public policymaking within the state. It is this group of women leaders—American Indian and Hispanic grassroots activists and public officials—who shape environmental policymaking in New Mexico and who are the focus of my study.

Despite their considerable involvement, however, relatively little research has been conducted into the political socialization, ideology, participation, or policy preferences of American Indian women and Hispanas in New Mexico, or elsewhere. Until recently, in fact, women of color were marginalized in the political science literature. Nevertheless, this situation is changing due in large part to the contributions of feminist scholars who are commited to documenting the political leadership and achievements of women of color. As more women and racial/ethnic minority scholars enter the field, we are likely to see new research into the political participation of these groups. This is certainly what has happened over the past 20 years to the study of gender politics.

As larger numbers of women have entered the academy, scholarship in women studies and feminist theory has grown and a substantial body of literature on women and politics has emerged. This work has produced important findings about women's involvement in electoral politics as elected and appointed officials, in grassroots politics as neighborhood activists, in labor unions as organizers, and in social movements as leaders. Fueled by debates among feminist scholars, this research has increasingly focused on the intersections of gender, race, and ethnicity and the role of these factors in women's political activism. As these studies further our knowledge of the political participation of racial/ethnic minority groups such as American Indian and Hispanic women, they help build the foundation necessary for developing theory. With this project, I also hope to contribute to this scholarship, (which I will review in greater detail later in this chapter). Before undertaking any further discussion of my study, however, I believe it is important to explain my position and motives as a researcher.

WHY STUDY THE POLITICS OF AMERICAN INDIAN WOMEN AND HISPANAS?

All inquiry is influenced, either directly or indirectly, by the worldview and theoretical assumptions of the researcher. Preferring to be clear about my own perspective, I will state my bias up front. I am concentrating on American Indian women and Hispanas for several reasons. First, we simply need to learn more about their political involvement because, as stated above, these groups have largely been ignored by the discipline. In particular, there is a dearth of scholarship on the politics of indigenous women.

Moreover, few studies include both grassroots activists and elected and appointed officials in their samples, and none (that I know of) have compared indigenous and Hispanic women leaders. As a result, the existing body of knowledge is still somewhat limited. More empirical research is needed to provide a solid basis for theory building. I see an opportunity to help with filling this gap by making a contribution to the literature. Furthermore, this project may generate some interest in the politics of American Indian women and Hispanas and, hopefully, will stimulate additional study.

My second reason for pursuing this research agenda is fundamental to pluralism and to our system of representative democracy. American Indians and Hispanics together make up 47 percent of New Mexico's population—a significant portion of the state's polity (U.S. Department of Commerce, Bureau of Census 1990a). These groups have a history of political activism in the state and currently exert considerable influence in New Mexico politics. What's more, American Indians and, in particular, Hispanics are a growing force in American national politics.

Third, I want to honor American Indian and Hispanic women leaders in New Mexico who work for social justice and environmental quality. They make valuable contributions to life in their communities and to the state, overall. Their stories have the potential to inspire others while adding to and enriching our collective history, the telling of which has traditionally excluded the experiences of women and racial/ethnic minorities. At a time when so many Americans are disengaged from politics and cynical about the motives and morals of our public officials, these women are exemplars of community leadership. Their paths to leadership, their policy goals, their practice of politics, and their strategies for influencing change provide us with models that we can emulate.

Fourth, in recent years, as their presence in formal (electoral) politics has grown, the role and influence of American Indian women and Hispanas in public policymaking has gained prominence. This has occurred despite their being "double minorities"—women and racial/ethnic minorities—both of which are marginalized by the dominant American culture and often lack significant political or economic power. It is valuable for both theory and praxis to learn how these women have overcome social and structural barriers to emerge as key players in the public policy process. I believe we can learn a great deal from their particular experiences and conceptualizations of politics. For example, the answers to questions such as "How and why did these individuals become politicized?," "How did they achieve positions of political leadership?," and "What strategies did they use to affect change?" can provide guidance for individuals and groups interested in building effective community leadership and developing political power.

My fifth reason for studying the politics of Indian women and Hispanas is personal and completely subjective. I am tremendously interested in learning more about the leadership of indigenous women and Hispanas in environmental policymaking because of who they are (women and minorities) and what they are doing (seeking social and environmental justice). No doubt, this has something to do with who I am: a Latina feminist, mother, and resident of the planet who is very much concerned with the increasing degradation of our earth's environment. This leads to the question "Why study environmental politics?" I have several reasons, which are outlined below.

WHY STUDY ENVIRONMENTAL POLITICS IN NEW MEXICO?

New Mexico is an ideal setting in which to conduct my research project because environmental policy is a highly salient topic here, many of the organizations linked to the environmental justice movement in New Mexico are active in state and local politics, and the presence of racial/ethnic minorities, especially American Indians and Hispanics, is fairly high.[3] The fact that environmental issues are especially salient in New Mexico can be explained by the state's history and by recent events. Since the 1940s, New Mexico has been a center for the development and testing of (first atomic and later) nuclear weapons by the United States Departments of Defense (DOD) and Energy (DOE). The excavation of uranium in the state has been a critical component of what is often referred to as "defense research." While DOD and DOE weapons programs provide employment and contribute to the economy of the state, they have produced a legacy of environmental and public health problems. These costs, some argue, far outweigh any benefits the state might receive from hosting such programs. In particular, Navajo and Pueblo peoples have historically borne the impact of often fatal health problems and environmental damage due to uranium mining on tribal lands. To a lesser extent, Hispanic communities around the state have also been adversely affected by large commercial mining operations that have harmed wildlife and resulted in significant soil and water pollution.

Wary of the trend towards siting locally undesirable land uses (LULUs) in poor communities and communities of color, residents have increasingly organized against such efforts. These grassroots groups have grown, building coalitions with other organizations and forming what has become known as the environmental justice movement. (I will discuss the movement in greater detail in chapter 3). In New Mexico, American Indian and

Hispanic women have been particularly active in this movement and comprise a significant portion of the leadership. In the early 1990s, for example, members of the Navajo Nation and the Mescalero Apache tribe successfully defeated proposals to place hazardous waste storage facilities on their tribal lands. In both cases, the grassroots campaigns were spearheaded by women. Also during this period, under the leadership of its first female governor, Isleta Pueblo adopted its own nationally recognized environmental standards in order to combat pollution generated by the City of Albuquerque. Similarly, neighborhood groups in Albuquerque's South Valley, a predominantly Hispanic area, pressured the city to locate the source of groundwater contamination. The investigation resulted in a number of wells being capped. When the pollution was found to emanate from the nearby weapons research laboratory located at Kirtland Air Force Base, community leaders, with the help of a local environmental justice organization, managed to negotiate the clean-up of several toxic sites.

In addition to these localized problems, New Mexico has, through an act of congress, become a recipient of much of the nation's medium- and low-level radioactive wastes. The DOE constructed and now oversees the controversial Waste Isolation Pilot Project (WIPP) in southeastern New Mexico, a storage facility for the radioactive wastes produced by various public and private entities around the country. As the waste is transported across the state to the WIPP site by trucks on public highways, concerns over safety and emergency responsiveness have been widespread. These examples illustrate some of the environmental problems that face communities throughout the state and contribute to the public's generally high level of interest in environmental issues. They also highlight the leadership role that American Indian and Hispanic women have played in the state's growing environmental justice movement. Having outlined the reasons for pursuing my research topic and the context for studying environmental politics in New Mexico, I will now describe the project in more detail.

THE PURPOSE AND SCOPE OF THE RESEARCH PROJECT

This study seeks to reveal some of the ways in which politics is conceptualized and practiced by indigenous and Hispanic women leaders in New Mexico. As women's politics comprises leadership in formal governmental institutions as well as in community based organizing, participants in this project include both public officials and grassroots activists from throughout the state. Specifically, I examine the motivations, ideology, policy priorities, and strategies for influencing the public decision-making process exhibited by native and Hispanic women leaders. The following questions guide the research.

1. What motivated these indigenous and Hispanic women to enter politics?
2. What are the political goals of these indigenous and Hispanic leaders?
3. What, if any, role do racial/ethnic identity and gender identity play in the formulation of their political goals?
4. What is their political ideology? Are they feminists? Environmentalists?
5. What are the public policy agendas of these leaders? And, how do the policy agendas of the grassroots activists and public officials compare?
6. What strategies do these indigenous and Hispanic leaders use to influence public policymaking?

While there is a very small nascent body of research on the political participation of native women and Hispanas in the United States, this literature does not examine their role as leaders in the arena of environmental politics. Nonetheless, this scholarship does provide an emerging theoretical foundation from which I developed a set of propositions corresponding to the six research questions. These propositions provide the organization for this project. A brief explanation and citations of the supporting literature are included in the discussion of each proposition, while a more comprehensive review of the relevant scholarship is presented at the end of the chapter.

The first proposition suggests that indigenous and Hispanic women, whether grassroots activists or public officials, will be similarly motivated to enter politics by a sense of civic obligation, by the opportunity to empower others, and by the desire to improve the quality of life in their racial/ethnic communities. Much of the research conducted on women in electoral politics has examined the motives of female elected and appointed officials for seeking public office. Many of these studies have concluded that women are largely motivated to enter politics by a sense of civic duty often acquired from their families, during their youth, as part of their political socialization (see for example Cantor and Bernay 1992; Fowlkes 1992; Le Veness and Sweeney 1987). In many cases, women in public office report having family members who are, or were, very active in politics (see Genovese 1993; Cantor and Bernay 1992). Related to their feelings of social responsibility is the goal of empowering others (to help themselves) and a concern for the well-being of individuals and their communities.

Researchers seldom examine electoral and grassroots politics together, assuming, perhaps, that these arenas and the political actors within them are too diverse for purposes of comparison. However, the few studies that do combine samples of female grassroots activists and public officials hint that there may be greater similarities among these actors than we might have previously thought, or than may exist among male activists and officials. For example, (maternal) feminist theorists argue that women are generally socialized to care for, or look after the needs, of others. Consequently, women tend to view politics as a web of human relationships through which shared goals may be accomplished cooperatively (see for example Sidel 1995; Tronto 1993; Tolleson Rinehart 1992; Phillips 1991; Elshtain 1981; Ruddick 1980). Indeed, scholarship addressing the role of women in social movements and grassroots politics worldwide has produced substantial evidence of women's community-based activism to improve social, economic, and environmental conditions (see for example Naples 1998b; Chowdhury and Nelson 1994a; Bystydzienski 1992a; Tilly and Gurin 1990; West and Blumberg 1990; Fincher and McQuillen 1989; Morgen and Bookman 1988). Furthermore, there is considerable empirical support for the above assertions in the small but growing literature on American Indian and Hispanic women's political participation, which leads to the following proposition.

Proposition two posits that indigenous and Hispanic women leaders, whether grassroots activists or public officials, will share at least two principal political goals. First, these leaders will seek to empower members of their racial/ethnic group and women in order to increase the political participation and influence of these groups in public policymaking. Second, the New Mexico leaders will aim to improve the overall quality of life for their racial/ethnic communities. As stated earlier, there is a burgeoning body of literature that examines the politics of women of color. Much of this research has shown that, regardless of the form of participation in which they engage, women of color generally conceptualize politics as a network of human relationships, that they value participatory democracy, and that they tend to view politics as a way of helping others (see Kaplan 1997; Hardy-Fanta 1993; Di Chiro 1992; Ford 1990; Pardo 1990). In effect, they use politics to both empower members of their racial/ethnic group and to improve the quality of life in their communities (see for example Pardo 1998; Sen 1995; Jaimes 1992; McCoy 1992; Albrecht and Brewer 1990). Research addressing women's political leadership, in general, has produced significant evidence of their efforts to promote women's policy concerns and to advance the position of women in society (see for example O'Regan 1997; Kahn 1996; Dodson et al. 1995; Thomas 1994; Tolleson Rinehart 1992; Darcy et al. 1987). On the other hand, critiques of feminist theory and of the women's movement have revealed the greater salience of racial/ethnic identity over gender identity for many women of color (see for example Pesquera and Segura 1993; Trask 1993; Johnson-Odim 1991; Garcia 1989; Thornton Dill 1987; Gunn Allen 1986; hooks 1981; Cotera 1980. See also Cohen et al. 1997; Junn 1997; Afshar 1996; Liebowitz and Carroll 1996; Basu, 1995; Baca Zinn and Thornton Dill 1994; Stack 1994). Based on this scholarship, we can assume that racial/ethnic identity and, perhaps to a lesser extent, gender identity will shape the political goals and policy agendas of American Indian women and Hispanas.

Proposition three asserts that racial/ethnic identity and gender identity will both play an important role in the formulation of indigenous and Hispanic leaders' political goals. However, racial/ethnic identity will be more salient for indigenous women than it will be for Hispanas while gender identity will be more salient for Hispanas than it will be for indigenous women. Similarly, racial/ethnic identity will be more salient for grassroots activists than public officials, while gender identity will be more salient for public officials than it will be for grassroots activists.

In recent years, feminist scholars have made significant inroads to our understanding of the role of both racial/ethnic identity and gender identity in women's political activism. Research on the political participation of native women and Hispanas has provided increasing support for the significance of race, ethnic identity, and gender to the formulation of their political goals and to their policy preferences (see for example Pardo 1998 and 1995; Hoikkala 1995; Cruz Takash 1993; Hardy-Fanta 1993; Jaimes 1992; McCoy 1992; Bonilla-Santiago 1991; Ford 1990). Racial/ethnic minority groups similarly seek social justice, influence in the political process, and greater economic equity.

While both American Indian and Hispanic people have struggled to overcome the oppressive social and economic conditions that their groups have historically faced in the United States, their experiences of colonization have been significantly different (see for example McClain and Stewart 1995; Takaki 1990, 1987; Blauner 1987. Also see Prindeville, et al. 1992; Stiffarm 1992; Jarvenpa 1989; Nelson and Tienda 1989; Segura 1987). Furthermore, as Native Americans have a unique legal and political status (they hold dual citizenship as members of sovereign Indian nations and as citizens of the United States) as well as their own tribal political systems, they are less likely than other minority groups to fully assimilate into American political culture (McClain and Stewart 1995). As a result, the willingness of native people to embrace dominant political ideologies or to identify with mainstream social movements will differ somewhat from that of other groups, such as Hispanics, for example. Based on differences between the degree of acculturation of American Indians and Hispanics, then, we can avticipate that differences will exist between the degree of racial/ethnic identity and gender identity expressed by the New Mexico leaders.

The literature on social justice movements suggests that activists seeking change for a particular racial/ethnic group are likely to share a strong identification with that group. As activists, they are able to embrace passionately a relatively narrow issue area or a particular interest group. On the other hand, a public official representing a broader constituency must, of necessity, consider the sometimes conflicting interests of a larger number of diverse groups. Hence, while public officials may exhibit a strong racial/ethnic identity, their political behavior will necessarily be tempered by their responsibility to represent a range of competing interests. Therefore, in general, both the degree and salience of racial/ethnic identity and gender identity will differ between grassroots activists and public officials.

The fourth proposition argues that while the political goals of indigenous and Hispanic women will be congruent, ideological differences will exist between these groups. In particular, indigenous leaders will be less likely than Hispanic leaders to self-identify as either feminists or environmentalists. Similarly, grassroots activists will be less supportive than public officials of mainstream feminist or environmentalist ideologies. Following from the argument outlined above, we can also assume that individuals holding leadership positions in the established political system have, for the most part, assimilated into the mainstream political culture and embrace conventional political ideologies (which are limited largely to the range between liberal and conservative versions of a representative democracy and a capitalist economy).

In contrast, "outsiders" actively challenging the status quo, such as grassroots activists, will most likely reject dominant political ideologies and even alternatives to these unless they represent the particular interests and goals of the activist and lead to significant change from the existing political system. To put this another way, we can assume that public officials are more likely than grassroots activists to embrace either feminism or environmentalism, ideologies born of social movements but which have now become incorporated, to varying degrees, into the dominant political culture. These ideological differences between grassroots activists and public officials will be similarly evidenced in their policy agendas.

Proposition five suggests that the public policy agendas of indigenous and Hispanic leaders will similarly reflect their goals of empowering members of their racial/ethnic group and women, and second, improving the overall quality of life for their racial/ethnic communities. However, the public policy agendas of grassroots activists will be somewhat narrower in scope than the broader agendas of the public officials. While only a few studies have explored the public policy preferences of Hispanic and American Indian women leaders (see for example Prindeville and Braley Gomez 1999; Bretting and Prindeville 1998; Prindeville and Bretting 1998; Knack 1995; Chiste 1994; Cruz Takash 1993; Hardy-Fanta 1993; McCoy, 1992; Bonilla-Santiago, 1991; Lynch, 1986; Willard 1984), the findings reveal that the leaders' policy agendas generally reflect their goals of empowering members of their racial/ethnic group and improving the overall quality of life for their communities. Building on the above discussion, we would expect that the leaders' goals would be embodied in a more or less narrow range of policies advocated by grassroots activists, and in a much broader array of policies promoted by public officials. The strategies used to achieve their goals may be more similar, however.

The sixth and final proposition posits that indigenous and Hispanic leaders will use similar strategies to influence public policymaking. Furthermore, whether they are grassroots activists or public officials, the leaders will employ both conventional and unconventional types of strategies for policy change. Political scientists tend to assume that social movement activists and public officials will have divergent policy agendas and that they will employ substantially different tactics for affecting change. Unfortunately, however, studies seldom compare the political behavior of grassroots activists with that of public officials, so these assumptions have yet to be fully tested. Nonetheless, based on various findings from the literature on women and politics, I wish to offer a different perspective.

Historically, women worldwide have used an array of diverse strategies to affect politics and policymaking (see for example Delamotte et al. 1997; Chowdhury and Nelson 1994a; Bystydzienski 1992a; Devens 1992; Ford 1990; Tilly and Gurin 1990; West and Blumberg 1990. Also see Bellows 1996; Brú-Bistuer 1996; Campbell 1996; Sen 1995; Albrecht and Brewer 1990; Hamilton 1989; Susser 1988). Traditionally excluded from the formal arena of institutionalized (electoral) politics, women have had to use whatever means were available to them and have, therefore, often relied on unconventional means for influencing policy change. In recent years, as formal barriers to women's participation have fallen, greater numbers of women have entered formal (electoral) politics. Many of these women have gained important leadership experience as grassroots activists in social movement politics. We might expect that they would bring with them unconventional approaches to politics and policy change. Therefore, based on women leaders' shared experiences of grassroots activism we can speculate that, regardless of their racial/ethnic identity or whether they are grassroots activists or public officials, women will employ some of the same types of unconventional strategies to effect policy change. And, building on the scholarship on women's political participation, we can further speculate that their practice of politics will be more congruent than dissimilar.

Of course, the only way to find out whether there is any support for the six propositions outlined above is to examine these in the context of the data, which brings us back to the purpose of my study. I believe that by comparing public officials with grassroots organizers relative to their experiences of politicization, their goals and political trajectories, we can better understand how and why American Indian and Hispanic women become political leaders, and what they seek to accomplish. Furthermore, we can identify their strategies for change, explore the effects of racial/ethnic identity and gender identity on their policy agendas, and relate them to their particular conceptualizations of politics, feminism, and environmentalism. In order to understand the role that these particular American Indian and Hispanic leaders play in New Mexico politics, however, it is instructive to review briefly the unique history of women in the region.

A BRIEF HISTORY OF WOMEN IN NEW MEXICO

Today, there are nineteen distinct sovereign Indian Pueblos in New Mexico: Acoma, Cochiti, Isleta, Jemez, Laguna, Nambe, Picuris, Pojoaque, Sandia, San Felipe, San Ildefonso, San Juan, Santa Ana, Santa Clara, Santo Domingo, Taos, Tesuque, Zia, and Zuni. The Pueblo people, comprising three linguistic groups: the Tanoans (consisting of the Tiwas, Tewas, and Towas), the Keresan, and the Zunis, are the original inhabitants of what is now called New Mexico. Archeological remains date their ancestors as living in the Southwest as far back as 10,000 B.C. (Sando 1992. Also see Brugge 1983; Eggan 1979; Ortiz 1979; Woodbury 1979). In contrast, the Athabascans—Apaches and Navajos—were relative newcomers, having migrated to the region around 1400 A.D. The Pueblos, the Apaches, and the Navajos were the principal groups residing in New Mexico when the Spanish arrived in 1539.

Women in PreColonial and Spanish Colonial New Mexico, 1582-1810

> ...The elders had no problems with me running for Governor. The elders had no problems with me going into the *kiva* because women are very powerful. It's been over the last 500 years or 300 years that

> they've been diminished by this mentality that our men took on from the Spaniards. [In] the true teachings of our people, the mother is the most important, the most sacred thing there is, which is the woman. So, as far as they were concerned, all I was doing was just bringing it back [full] circle. So they supported me (Valerie, public official).

Prior to European contact, the aboriginal people of Southwestern North America lived in bands as hunters-gatherers and in horticultural or agricultural communities distinguished largely by female-male reciprocity and complementarity (Sando 1992; Gutierrez 1991; Albers 1989; Gunn Allen 1986. Also see Ortiz 1979; Woodbury 1979).[4] Individuals were ascribed gender specific roles and responsibilities, which in many cases were based on egalitarian norms that included shared decision-making (see Klein and Ackerman 1995; Devens 1992; Albers 1989 and 1983; Shapiro 1986; Shepardson 1983; Medicine 1978; Niethammer 1977). While the lifestyles of these tribes varied—Athabascans were hunters-gatherers and Pueblos were agriculturalists—their cultures were similarly characterized by social and religious systems in which a female figure, such as corn mother or spider woman, held a central position (Sando 1992; Gutierrez 1991; Gunn Allen 1986). Furthermore, many of these societies were matrilineal and/or matrilocal and these arrangements may have facilitated women's holding valued economic, religious, and/or social positions within their tribe[5] (Jaimes 1992; Albers 1989 and 1983; Bonvillain 1989; Eggan 1983 and 1979; Etienne and Leacock 1980; Medicine 1978). In the case of Pueblo society, these female-centered systems contributed to women having political clout and personal autonomy, and to their exercising authority even where their sex was excluded from holding formal governmental positions such as that of War Chief or Medicine Man[6] (Jacobs 1995).

In the sixteenth century, Spanish conquest of the area they christened *Nuevo México* led to significant changes in the lives of the native residents. The goal of the *conquistadores* [conquerors], like that of the English on the east coast of North America, was to expand the crown's territories for the purpose of enriching Spain's influence and wealth as a European power.[7] The plunder of New Spain's natural resources was facilitated by the exploitation of the area's inhabitants through their simultaneous enslavement and forced conversion to Catholicism.[8] Gradually, these factors contributed to the systematic corruption of the cultural traditions, social practices, and religious belief systems of the peoples of the Southwest (Jaimes 1992; Sando 1992; Gutierrez 1991; Albers 1989 and 1983; Gunn Allen 1986; Eggan 1979; Simmons 1979a). Patriarchal institutions and social systems supplanted those previously indigenous forms. The social and economic transformation from egalitarian or ranked relations of production to preindustrial stratified societies had far-reaching consequences for native communities resulting in the degredation of the role and status of aboriginal women in particular (Chiste 1994; Devens 1992; Albers 1989; Bonvillain 1989; Etienne and Leacock 1980).[9] As one Native scholar explains,

> When you devastate a traditional subsistence economy, you force women from a position of being equal into a position of being marginalized parts of the dominant society. ...So it is an important issue to look from the standpoint of the rights of women to have some control over their lives, and to have some control over their future, and the right to be empowered in their work (LaDuke 1991, 17).

In spite of the hostile conditions imposed by Spanish rule and Catholicism, Indian women fought for, and in some cases succeeded in maintaining, a central place in their religious belief systems if not in the political life of their communities[10] (Jacobs 1995; Shepardson 1995; Sando 1992; Gutierrez 1991; Gunn Allen 1986).

Grants of (Indian) land or *mercedes* were initially awarded by the crown to individual *conquistadores* to facilitate colonization of New Mexico. The *haciendas* [large ranching estates] of these territorial administrators, together with the Franciscan missions, formed the nucleus of the early Spanish/Mexican settlements.[11] Later in the eighteenth century, land grants were generally made to familial groups of settlers who held the property in common for farming and grazing their herds. This strategy aided colonization by establishing communities made up of large extended families who shared their resources for reasons of economy, safety, and custom.

In both aboriginal and Spanish/Mexican villages, women assumed primary responsibility for socializing and instructing young children, caring for family members and their livestock, for ministering to the sick and aged, for building and maintaining the home, preparing the family's food, and making the family's clothes as well as household utensils and other tools (Rebolledo et al. 1992. Also see Forrest 1989; Kutsche and Van Ness 1981). In general, men and women held different roles and exercised varying degrees of authority in their respective public and private domains. Where families found themselves struggling to survive, the interdependence necessitated by an agropastoral economy and dangerous frontier conditions sometimes blurred the division of labor between the sexes.[12] As needed and when practicable, women and men shared activities such as childcare, home maintenance, the manufacture of household goods, and food production (Jacobs 1995; Niethammer 1977; Gonzalez 1969).

Furthermore, on occasion, both women and men, whether Indians or *pobladores/as* [colonists], lost their lives in defense of their families, homes, fields, and livestock during attacks by Navajo, Apache, or Comanche raiders (Rebolledo et al. 1992; Gutierrez 1991; Forrest 1989; Gunn Allen 1986; Brugge 1983; Lange 1979). Overall, however, in both the family and in the larger community, Hispanic men enjoyed considerably greater autonomy and decision-making power than Hispanic women (Rebolledo et al. 1992; Forrest 1989; Kutsche and Van Ness 1981; Gonzalez 1969).

Many of the tasks traditionally assigned by sex in Indian communities were altered by the Fransiscan missionaries intent on imposing their patriarchal values. In pre-colonial native societies the nature of one's work was frequently determined by one's sex so that, for example, while women engaged in homebuilding, weaving, and pottery-making, men participated in hunting and agricultural activities.[13] While these tasks defined the gendered roles of women and men, they served to maintain a complementarity between the sexes. As their roles were modified, and the balance between women and men's access to and control of resources was upset, the status and autonomy of women was affected negatively (Albers 1989 and 1983; Gunn Allen 1986).[14] Furthermore, native social structures that generally valued knowledge, skill, and maturity—qualifications obtainable by both sexes—were replaced by a rigid colonial hierarchy based on race, ethnicity, ancestry, occupation, ownership of land, religion, and sex (Gutierrez 1991; Gunn Allen 1986). Nevertheless, because many of the Southwestern tribes traditionally placed a greater value on women and their social contributions, aboriginal women enjoyed greater autonomy, security, and (at least in their tribes) higher status than their Spanish or *mestiza* [mixed blood] counterparts (Klein and Ackerman 1995; Shepardson 1995; Maltz and Archambault 1995; Jaimes 1992; Niethammer 1977).

Compared with Pueblo or Navajo women, Spanish women and mestizas faced more social restrictions and had fewer available life choices (Rebolledo et al. 1992; Gutierrez 1991; Kutsche and Van Ness 1981; Gonzalez 1969). Women living in colonial settlements in New Mexico were subject to a patriarchal culture obsessed with masculinity and perverse notions of honor, which maintained double standards of conduct and morality for women and men. However, those most constrained by the controls on women's social and sexual behavior were Spanish women and mestizas from the upper classes (Gutierrez 1991). While affluent families could "protect" the honor of their women by largely confining them within the walls of their *haciendas*, most *pobladores* could not meet the demands of life on their farms or ranches without the contributions of women's labor. As a result, women of the working classes were, of necessity, somewhat less hampered by social custom. In fact, many *pobladoras* were widows or single women who ranched, farmed, and/or traded their crafts and homemade goods to support their extended families (Rebolledo et al. 1992; Gutierrez 1991). The harsh living conditions engendered cooperation and interdependence among women who faced similar adversities and who struggled to care for their families. As a result, these women negotiated their lives using their strength and determination, their faith, their influence, and when necessary, the legal rights available to them under Spanish law.[15]

During the approximately 250 years of Spanish occupation, numerous significant economic and social changes ensued, which resulted in the intermingling of the aboriginal, mestizo, and Spanish cultures. These changes included the imposition of Catholicism and patriarchy on native peoples, the pervasive enslavement of Indians as domestic servants and agricultural workers, the exchange and adaptation of agricultural products and technologies between natives and *pobladores*, the transformation of local economies, and finally, the growing *mestizaje* [mixing of the Indian and Spanish races] of the population resulting from intermarriage and the widespread rape of both *genízaras* and free Indian women[16] (Rebolledo et al. 1992; Gutierrez 1991; Chavez 1979; Eggan 1979; Gonzalez 1969). These factors contributed to the complex mix of racial, ethnic, and cultural relations characteristic of New Mexico, which carried into the Mexican period (1821 to 1846) and beyond.[17]

Women in Mexican and Territorial New Mexico, 1810-1911

By the time that Mexico gained its independence from Spain in 1821, a uniquely Southwestern cultural mosaic had emerged in New Mexico. Control over the region, however, shortly passed to the United States when in 1848, as a result of the Mexican-American war, Mexico surrendered the territory with the signing of the Treaty of Guadalupe Hidalgo. American colonization soon prompted dramatic new social, economic, and political changes for the inhabitants of New Mexico (Forrest 1989; Rodriguez 1987; Sierra 1984; Brugge 1983; Roessel 1983; Kutsche and Van Ness 1981; Simmons 1979b). While these radical changes adversely affected Spanish/Mexican and Indian communities to varying degrees, the resulting social and economic upheaval particularly affected the economic position of women. For example, due to the strict trade embargo levied by Spain on the New World, women had produced numerous goods for local consumption for many years. Similarly, during the period of struggle for Mexican independence, the territory was cut off from nearly all legally supplied commodities. Women's productive labor had contributed in important ways to the barter economy providing women with both resources and greater autonomy (Knack 1995; Rebolledo et al. 1992; Campos Carr 1988-89). The opening of the area to commerce and the introduction of a cash economy by the United States ushered in significant changes to the lives of women—providing new economic opportunities which were, however, limited to increasingly menial work.

For the Indian tribes of New Mexico, colonization once more meant an assault on native people, their culture, religious beliefs, livelihood, customs, language, and way of life. Upon gaining control of the region, the United States sought to assimilate the native population by displacing many of the tribes to reservations,[18] reducing severely their land base and indigenous food sources, and removing Indian children to government and mission boarding schools,[19] forbidding the speaking of aboriginal languages, and prohibiting Indian people their traditional religious practices (Churchill 1992; Devens 1992; Robbins 1992; Sando 1992; Stiffarm 1992; Takaki 1990; Gunn Allen 1986; Brown 1970).

Non-Indians, formerly Spanish and later Mexican citizens, were now Mexican-Americans. They too were targets of the American government's policy of forcible assimilation. They too lost many of their political and social institutions, their local economies were transformed, much of their land was taken, families were displaced, women's legal rights were curtailed, and children were discouraged or prohibited from speaking Spanish in the schools (Kelley 1999; Rebolledo et al. 1992; Takaki 1990; Forrest 1989; de la Garza and Vaughan 1985; Sierra 1984; Bayes 1982). However, the Treaty of Guadalupe Hidalgo established some protections for both Indians and Mexicans, including the right to preserve their religion, culture, and language (see especially Sando 1992; Brugge 1983; Roessel 1983; Simmons 1979a and 1979b). These guarantees aided New Mexico's Hispanic leaders in maintaining their power under the new regime to the extent that they played an important role in the creation of the state's political system (Rebolledo et al. 1992; de la Garza and Vaughan 1985; Sierra 1984). Consequently, the Hispano elites who controlled the political infrastructure established by the Spanish, and in existence when the region became a U.S. territory, were able to influence New Mexico's path to statehood.

Despite the formidable pressures placed on the territory's Hispanic leadership to change and adopt Anglo American systems, their role in the process helped to protect Spanish culture and to retain a space for Hispanos' political participation in the new order. The majority of the framers of the New Mexico state constitution were Hispanos. They led the development of new social and political organizations while ensuring that the constitution contained the following provision:

> The right of any citizen of the state to vote, occupy office or be a member of a jury shall never be limited or forbidden for reasons related to religion, race, language or color, or the inability to speak, read or write the English or Spanish languages... (Article VII, Section 3, *Constitution of the State of New Mexico*).

While Hispanic males were able to preserve a place for themselves within the new American political system, Indian men and essentially all women were disenfranchised—even though they remained the majority population in the state into the twentieth century (de la Garza and Vaughan 1985). For example, although Indian males could participate legally in federal elections they were prohibited from voting in state races. On the other hand, all women in New Mexico were denied suffrage in both state and national elections, and were largely excluded from any involvement in electoral politics. Nevertheless, these barriers to participation in public life would soon be torn down by the very groups that they sought to exclude.

Women in Twentieth Century New Mexico, 1911-Present

> While (American Indian) women still play the traditional role of housekeeper, childbearer, and nurturer, they no longer enjoy the unquestioned positions of power, respect, and decision making on local and international levels that were not so long ago their accustomed functions. Only in some tribes do they still enjoy the medicine or shamanistic power they earlier possessed. ...[W]ith the coming of the white man and his patriarchal system, the powers of the women were systematically undermined in countless ways, and this undermining was and is reinforced willingly by many of the men (Gunn Allen 1986, 202-203).

While New Mexico's Hispanos and Indians preserved important elements of their culture, including their language and religion, the new social order privileged Anglo Americans over other racial/ethnic groups (Maltz and Archambault 1995; Jaimes 1992; Forrest 1989; Gunn Allen 1986; de la Garza and Vaughan 1985; Sierra 1984; Kutsche and Van Ness 1981; Cotera 1980). At the same time that racial prejudices were increasing, non-Indian women were beginning to gain some of the rights that men enjoyed under the new state government. For example, in the process of creating New Mexico's Constitution, negotiations between Hispano and Anglo political elites resulted in both Anglo and Hispanic women gaining the right to vote in school board elections (Rebolledo et al. 1992). A decade later, following ratification of the Nineteenth Amendment to the U.S. Constitution, which gave American women the right to vote, New Mexico voters passed the Elective Franchise Act. While women had been eligible for certain political appointments within the territorial government prior to statehood, this Act allowed them to run for all elected offices in the state[20] (Belmonte 1996; Rebolledo et al. 1992). Consequently, women ran for and won numerous state and local races for public office (see Robertson 1995).

Like their Anglo American counterparts, however, most Indian and Hispanic women were only peripherally involved in formal electoral politics. They cared for their families, participated in community life, and struggled to sustain their traditions in changing times (Rebolledo et al. 1992; Campos Carr 1988-89; Forrest 1989; Bataille and Sands 1984; Kutsche and Van Ness 1981; Cotera 1980; Elassar et al. 1980). They survived and prospered by adapting to their environment, taking on new roles, and creating their own mutual aid societies, labor, and civic organizations (Rebolledo et al. 1992; Cotera 1980; Elassar et al. 1980).[21] They used these associations to pressure the major political parties and governmental institutions to respond to their communities' needs. However, even as women won the vote and their political participation gained legitimacy, the role of Hispanos/as in the state's political life declined, and their influence on New Mexico's political and economic power structures decreased (Forrest 1989; Sierra 1984; Holmes 1967).[22] As Hispanos/as lost their proportional share of key political offices to Anglo contenders in the early decades of the twentieth century, their voice in state policymaking grew weaker and their frustration increased.

Even as New Mexico's Hispanos/as were struggling with the changes wrought by American colonization, native peoples, in particular, faced a long, hard battle to secure even basic civil rights (see McClain and Stewart 1995; Churchill 1992; Robbins 1992; Brown 1970). They were denied continually the right to participate in the politics of the state. For example, although Congress made all Indians citizens of the United States in 1924, New Mexico, Arizona, and Utah continued to deny American Indians voting rights. These states claimed that since Indian tribes and reservations were subject to federal jurisdiction, Indians were not citizens of the state and, therefore, were not eligible to vote in state and local elections. It was not until 1948, after lengthy litigation, that Indians finally won the franchise within these states.

In 1934, the U.S. Congress passed the Indian Reorganization Act (IRA) under which tribes were induced to organize with written constitutions and charters of incorporation such as those granted to business enterprises. The major purpose of this legislation was to place native tribes more directly under the control of the Bureau of Indian Affairs. The IRA, also known as the "Wheeler-Howard Act," compelled Indian nations to adopt an organizational structure approved by the U.S. government but contrary to traditional forms of native governance which were largely participatory in nature, generally inclusive of both sexes, and frequently based on consensus decision-making. In effect, the federally sanctioned "reforms" imposed representative democracy upon tribes including adoption of a constitution and election of a formal leadership. Ironically, this new system actually *limited* the political participation of tribal members, who, in many cases, had previously enjoyed a greater role in self-governance. In particular, women from traditionally egalitarian tribes found their political influence diminished. Native scholars argue that the IRA was "designed to undercut the unity marking traditional native societies, replacing it with a permanent divisiveness," and that "grassroots native resistance to the law was immediate, outspoken, and sustained" (Robbins 1992, 95. See also Churchill and Morris 1992; Sando 1992).

Despite the significant cultural, social, and political changes that resulted from the IRA, tribes willing to comply with the federal requirements gained certain advantages. A complete reversal of previous U.S. government policy, the IRA provided for the return of unsold allotted lands to tribes,[23] established day schools on reservations instead of off-reservation boarding schools, encouraged the practice of traditional cultural activities, and recognized the authority of tribal governments promoting both self-determination and economic development. Amendments to the Indian Reorganization Act have included provisions that allow certain rights of home rule for Indian nations, the conservation and development of Indian lands and resources, the right of Indian nations to form business and other organizations, establishment of a credit system for Indians, and provision of vocational education for Indians.[24] While these measures returned a degree of autonomy to Indian nations, the loss of their traditional forms of governance created an imbalance in the distribution of power within many tribes, further altering their indigenous cultures.

There is substantial evidence to support the claims of American Indian scholars that the assimilation of indigenous peoples into the dominant white society has remained the ultimate policy goal of the U.S. government. Nevertheless, the federal government's efforts to achieve this end have remained largely unsuccessful; especially where New Mexico tribes are concerned. Their ability to adapt and endure in the face of persistent and overwhelming adversity is evidenced by the fact that the Navajo Nation is now the largest tribe in the United States and that nineteen Pueblos continue to function as sovereign nations today. Furthermore, tremendous variation exists with regard to both the forms of governance and the political practices of tribes within the state. While some Indian nations adopted elements of the IRA, many did not.[25] For example, the Navajo Nation, which established a tribal council form of government during the 1920s, rejected the IRA (Shepardson, 1983). Most Pueblos continue using their traditional theocratic forms of government and/or retain significant features of the Spanish colonial system of governance.[26] At the same time, these Indian nations tend to restrict substantially who can participate in tribal politics and to what degree.

In many of the so-called "traditional" Pueblos, the civil rights and civil liberties guaranteed to citizens of the United States are not recognized by, nor are they extended to, tribal members. While these limitations on the rights of individuals affect both women and men, they especially impact women (see especially Aks 1996). For example, the majority of New Mexico's 19 sovereign Pueblos currently prohibit enrolled female members from participating in tribal politics. Of the 19 Pueblos, approximately six have written constitutions and about seven allow for varying degrees of women's participation in tribal government.[27] A smaller number currently allow women to hold political office including such policymaking positions as governor, lieutenant governor, and councilmember (see Linthicum 1998; Contreras and Shaw 1993). In most Pueblos, however, women are prohibited from attending tribal council meetings, from voting (where tribal elections exist), and from holding tribal office (see for example Linthicum 1996). Their disenfranchisement has discouraged many Indian women from seeking public office, either at the tribal or non-tribal level. Other women, such as some of the native leaders interviewed for this project, have found opportunities for political participation and civic involvement outside of their respective Pueblos in school boards, counties, state and/or national government.

When compared to the Pueblos, the Navajo Nation has afforded women the right to participate in tribal government to a greater degree and for a longer period[28] (see Eggan 1983; Roessel 1983; Shepardson 1983; Simmons 1979b). For example, as early as the 1950s, women were serving on the tribal council, the tribe's legislative body. Annie Dodge Wauneka (the first woman elected to the tribal council) and Irene Stewart (later elected to council) were influential leaders who advocated for women and children, and worked to improve housing, education, public services, and public health for the Diné people (Stewart 1980; Nelson 1972). Women have also participated extensively in chapter politics vocalizing their concerns, holding elected positions, and administering programs for their communities.[29]

Overall, however, the number of Indian women in tribal politics throughout the country remains low. As late as 1990, the U.S. Department of Interior, Bureau of Indian Affairs, reported that there were only 61 female tribal leaders for more than 500 tribes (McCoy 1992). Despite the greater inclusiveness of tribes such as the Navajo, women continue to face the same barriers and impediments that women generally do in American politics.[30] And yet, like their non-Indian counterparts, Indian and Hispanic women make important contributions to their communities (Prindeville and Bretting 1998). As one scholar notes,

> When viewing the creative role of women in the simultaneous processes of social change and community survival, one must conclude that if it wasn't for the women, racially oppressed communities would not have the institutions, organizations, strategies, and ethics to maintain itself (sic) as an integral whole, but also to develop in an alien, hostile, oppressive situation and to challenge it. In spite of their powerlessness...women of color generally have a dramatic impact within and beyond their communities (Gilkes 1994, 242).

In order to place this project within the context of the existing research on women and politics, we now turn to a discussion of the relevant literature.

WOMEN'S POLITICAL PARTICIPATION IN AMERICA

Historically excluded from fully exercising their rights of citizenship, American women have struggled to make themselves heard in the political arena. Despite the various legal, economic, and social barriers to their full participation, women—whether Hispanas, American Indians, or Anglos—have long engaged in pressure politics, grassroots mobilization, and other forms of political expression. Before gaining the franchise and up to the early 1960s, women were largely limited to either working behind the scenes or outside of the formal arena of electoral politics. Nevertheless, they had an impact on public policymaking by organizing around concerns affecting the family, public health and safety, and morals. With the rise of the social movements of the 1960s and 1970s, women burst onto the political scene in ever-increasing numbers. They mobilized over a broad range of issues including civil rights, women's rights, workers' rights, peace, the environment, gay and lesbian rights, and the rights of the disabled as well as for a host of other causes. These issues form the core of what West and Blumberg call "the politics of survival."

> Whether we consider women fighting for food, shelter, and clothing, or for jobs, or for the right to live as a national or racial/ethnic group with the same life chances as any other, or striving to minimize the effects of 'man-made' disasters, such as war and global pollution, or natural catastrophes, such as earthquakes or floods, or pursuing women's liberation struggles around the world, some common themes emerge. Women join the fight for their survival as women in many devalued roles in patriarchal society: mothers, workers, clients, students, lesbians, elderly, and second class citizens. ...The politics of survival...[is] a politics of love and caring that has propelled women to break out of their boundaries and to confront or enlist the forces of the state to change the balance of power (West and Blumberg 1990, 14-15).

While traditional women's issues are seen as pertaining to reproduction, nurturance, children, and health—reflecting woman's role in the domestic sphere—women's activism also extends into the workplace and the larger community (Orleck 1997; Basu 1995; Sen 1995; Tronto 1991; Fincher and McQuillen 1989; Morgen and Bookman 1988; Ruddick 1980). As social reformers, community organizers, and labor leaders in the nineteenth and early twentieth centuries, women championed social justice and agitated for public policies relating to abolition, temperance, suffrage, child labor-, orphanage-, prison- and mental institutional reforms, family planning, and unions (McGlen and O'Connor 1995; Evans 1993; Tilly and Gurin 1990; West and Blumberg 1990; Darcy et al. 1987; Flammang 1984a).

Though not yet equal partners in the game of politics, women in the United States have begun to play a more influential role in the public policymaking process. In the latter part of the twentieth century, women spearheaded the civil rights, women's, peace, and environmental justice movements and were instrumental to the labor, farm workers', Chicano, American Indian, Black Power, and gay and lesbian rights movements. One scholar notes that

> the issues forming the subject of women's activism, while practically limitless, are overwhelmingly about improving people's capacities to participate in society, about affirming life, and about getting things done far from the center of power—but with an eye on it (Milroy 1991, 8).

By establishing that the personal is political, the contemporary women's movement, in particular, legitimized many issues formerly seen as lying outside of the "proper" scope of politics and located them firmly on the public policy agenda.[31] Over the past 30 years, for example, due in large part to the groundbreaking work of the women's movement, women have succeeded in establishing public policies relative to child care, education, job training, employment, divorce, insurance, and welfare resulting in substantial gains to women's economic and legal status (O'Regan 1997; Dodson et al. 1996; Thomas 1994; West and Blumberg 1990; Saint-Germain 1989; Boneparth and Stoper 1988; Susser 1988).

Furthermore, women successfully placed a variety of issues on the public policy agenda that had long been taboo subjects such as domestic violence, sexual assault, incest, and abortion. These were problems that affected women across all classes, races, and ethnic groups. In broadening the public debate, the women's movement succeeded in making public policy more responsive to women's needs and a growing base of support was established from which Hispanas, Indian women, and others could voice their demands for equity and social justice. As a result, politics became both more representative of racially and ethnically diverse women's concerns and more relevant to women's gendered experiences. Furthermore, these changes influenced definitions of politics, the rules of the game, and who could play, resulting in greater opportunities for political participation for all women.

Women's increased influence on public policy today is largely a result of their greater political participation due in part to their involvement in, and changes stemming from, the women's movement and other social movements of the 1960s and 1970s. Like the feminist associations formed for the purpose of solving social ills and reforming public institutions during the nineteenth century, the organizations that emerged during feminism's second wave provided a fertile training ground for women interested in political activism (Ferree and Martin 1995; McGlen and O'Connor 1995; Evans 1993; Tilly and Gurin 1990; S.J. Carroll 1989b; Sapiro 1983).[32] As Basu notes,

> women's movements have increasingly sought to influence state policy and increase women's political representation and participation. In Mexico, Brazil, Chile, and the United States, women's movements have moved off the streets and into political institutions (1995, 16).

Similarly, because they addressed issues of nationalism, cultural domination, racism, class inequalities, and social injustice, the Chicano and American Indian movements also attracted many Hispanic and native women. In general, women learned valuable organizing skills, developed their political ideology, and gained confidence through their work in these various social and political movements.

The modern women's movement, in particular, succeeded in making tremendous inroads for women and led to the founding of several politically prominent organizations including the National Organization of Women (NOW) and the National Abortion Rights Action League (NARAL). Their representation of women's policy concerns in the political arena led to the passage of legislation and to judicial decisions that established important rights for women. However, while many women of color benefited from these gains and some were even active within the movement, they generally remained distant from the primarily middle-class Euro-American women's movement. Despite all that it accomplished to empower women, the women's movement largely failed to gain the support of the majority of poor and working-class women, lesbians, or women belonging to racial and ethnic minorities.[33] In fact, the women's movement was criticized for being classist, homophobic, racist, and largely unresponsive to the needs of poor women and to women of color. The concerns of these groups relative to economic equity, social justice, and racism fell outside of the political agenda set by the mainstream women's movement. While they built upon the achievements of the women's movement, women within the Chicano, American Indian, and Black Power movements, for example, developed their own conceptualizations of feminism, established their own organizations to meet their communities' needs, and formulated their own agendas for change. More importantly, these women became part of a pool of experienced and knowledgeable leaders who could be drawn on to run for public office or to spearhead an issue campaign.

Since 1992, women have been elected to national, state, and local level offices in record numbers. However, in proportion to men, the number of women in public office remains quite small (Fox 1997; Thomas 1997; Kahn 1996; Dodson et al. 1995; McGlen and O'Connor 1995; Burrell 1994; S.J. Carroll 1994). In contrast to women's negligible but growing presence in electoral politics, women's historical leadership in grassroots community organizing continues to provide one of the greatest opportunities for women to shape public policy (Rocheleau et al. 1996; Basu 1995; Ferree and Martin 1995; Bystydzienski 1992a and 1992b; Fincher and McQuillen 1989; Ackelsberg 1988; Morgen and Bookman 1988). We now turn to an examination of the literature on women in electoral and grassroots politics.

POLITICAL SCIENCE AND THE POLITICS OF AMERICAN INDIAN AND HISPANIC WOMEN

Feminist scholarship exploring the role of women in politics has flourished in recent years, especially during the decade of the nineties. Social scientists are now studying the political leadership of women in community-based organizing as well as in electoral politics. Several of these works are path-breaking for revealing the tremendous scope and diversity of women's social and political activism around the globe, and for laying important groundwork from which we can begin to build theory. Within this larger literature are critical studies, important for their analyses of the role of women of color as organizers and leaders of the growing environmental justice movement in the United States and abroad (see especially Kaplan 1997; Rocheleau et al. 1996; Peterson and Runyan 1993. See also Bretting and Prindeville 1998; Prindeville and Bretting 1998; Gedicks 1993; Hamilton 1993; Di Chiro 1992). A growing web of grassroots groups, founded and led by women, comprise the movement. In contrast to the established Euro-American environmentalist or "Green" organizations, the environmental justice movement is largely led by and composed of working-class people and members of racial/ethnic minorities. Their local organizations build on the goals of the civil rights and women's movements integrating issues of social- and racial justice with economic and gender equity to transform social and political institutions and environmental conditions. (For more on the environmental justice movement, see chapter 3).

This growing area of research examines the impact of gender, race/ethnicity, class, and the effects of globalization on women's environmental politics and policymaking. Most importantly, the findings indicate that women's experiences of politicization, how they conceptualize politics and power, the ways that they participate in politics, and even the effects that public policy has on women and their children, are often markedly different from men's experience (see for example Delamotte et al. 1997; Duerst-Lahti and Kelly 1995; Hardy-Fanta 1993; Peterson and Runyan 1993). Obviously, the implications are significant for the practice of a fully representative democratic politics. Much of this scholarship informs our understanding of women's political leadership in New Mexico's environmental policy arena. It is especially useful, therefore, to highlight the research findings within this developing literature that are most relevant to the study of indigenous Southwestern and Hispanic women's environmental politics.

New Scholarship on Women's Leadership in Grassroots Organizing

During the 1990s, a significant body of work was published examining the varied roles, ideologies, and multiple identities of women active in grassroots organizing. This research reveals that women's participation in community-based activism is motivated, shaped, and informed by women's roles and identities as mothers, providers, community residents, members of racial and/or ethnic minority groups, or by a combination of these identities (see for example Naples 1998b; Orleck 1997; Afshar 1996; Randall 1995; Bystydzienski 1992a; Tilly and Gurin 1990). As mothers, women's traditional reproductive and caregiving responsibilities frequently place them at the center of domestic evironmental health and safety and micro-economic concerns.[34] However, while motherhood often legitimizes and sanctions women's environmental advocacy, it may also limit the scope of women's activism and their ability to effect change outside of the narrow confines of such so-called "women's" issues[35] (Krauss 1998; Bellows 1996; Miller et al. 1996; Seager 1996; Pardo 1990. Also see Dietz 1985).

In a similar way, women's roles as nurturers and providers (with specific responsibility for the maintenance of the household/family unit) have simultaneously precipitated, informed, enabled, and constrained women's collective environmental action (see for example Kaplan 1997; Rocheleau et al. 1996; Wangari et al. 1996; Wastl-Walter 1996). Likewise, women's membership in and identification with a particular clan, tribe, or community; socioeconomic class or caste; and/or racial or ethnic group may either facilitate opportunities for women's organizing activities (see Pardo 1998; Brú-Bistuer 1996; Sen 1995; Hardy-Fanta 1993; Pulido 1993) or deter their involvement (see also Kaplan 1997; Campbell 1996; Mehta 1996; Shields et al. 1996) depending on existing legal, structural, cultural, and/or social barriers to women's political participation. Thomas-Slayter, Wangari, and Rocheleau point out, however, that in spite of the many obstacles they face, "...women find means of expression and rebellion, to engage in oppositional practice and to protest a range of injustices based on selective access to or degradation of resources for production, survival, and cultural continuity" (1996, 303).

Other scholars argue that over time and across cultures, women have vocalized their social, political, and economic concerns and taken action against injustice using whatever means available to them. West and Blumberg characterize the issues that have incited women's activism as falling into four general categories: "(1) those directly linked to economic survival; (2) those related to nationalist and racial/ethnic struggles; (3) those addressing broad humanistic/nurturing problems; and (4) those identified in different eras as 'women's rights' issues" (1990, 13). These researchers illustrate how women's movements around the world have, in recent years, incorporated peace and environmental causes into their agendas, uniting women globally in a politics of survival.[36] In combination, this scholarship reveals important parallels among the issues and strategies employed by women across cultures and, therefore, helps to inform our understanding of the politics of racial/ethnic minority women in the United States (see for example Delamotte et al. 1997; Kaplan 1997; Afshar 1996; Rocheleau et al. 1996; Randall 1995; Chowdhury and Nelson 1994a). In this way, scholarship addressing the motives, goals, and behavior of women involved in social movements and grassroots organizations illustrates the tremendous scope, influence, and creativity of women's collective action.[37]

Another feminist scholar, whose work informs my own research examines American women's diverse experiences of politicization, their motives for participation, their policy agendas, and their paths to political power. In her study of white women activists and public officials, Fowlkes (1992) reveals the importance of women's nontraditional sex role socialization—what she terms their "countersocialization"—to both the development of their political ideology (whether left or right in orientation), and to the form of activism that they practice (whether conventional or "radical"). The women she interviewed chose movement or party politics based on their beliefs about the sources of social problems and the ways to best solve them. Fowlkes also found that women's understanding of what it means to "be a woman" grounds their feminist or anti-feminist consciousness, shapes their definition of politics, and ultimately guides their participation in politics (1992). My own study similarly explores leaders' notions of feminism and whether their identification with other women as a group—their "gender consciousness"—informs their political goals or policy agendas.

Other scholars offer new ways of interrogating the social construction of gender in America by proposing an alternative feminism that incorporates racial/ethnic and economic analyses. Through a compilation of essays by and about women of color in the United States, Baca Zinn and Thornton Dill documented how women of color are subjugated socially, culturally, and economically because "patterns of hierarchy, domination, and oppression based on race, class, gender, and sexual orientation are built into the structure of our society" (1994, 4). Additionally, this research reveals how women's participation, opportunities, and access to resources and power is limited in numerous ways by social structures and institutions[38] (For example, see Joe and Miller 1994; Segura 1994. See also Naples 1998b; Rocheleau et al. 1996; Pesquera and Segura 1993; Peterson and Runyan 1993; Nelson and Tienda 1989; Morgen and Bookman 1988). This analysis is informed and guided by what Baca Zinn and Thornton Dill term "multiracial feminism,"

> ...a body of knowledge situating women and men in multiple systems of domination. ...[I]t treats racial inequality as a vital shaper of women's and men's lives and advances a coherent and powerful premise—that racial ancestry, ethnic heritage, and economic status are as important as gender for analyzing the social construction of women and men (1994, 11).

As the global political economy increasingly affects their status and choices, women are building coalitions across divisions of race/ethnicity, class, and geographic region to collectively influence public policy and work for social change (Albrecht and Brewer 1990). In combination, these analyses reveal the connections among women's multiple roles, "environmental concerns, local knowledge, gender-based responsibilities and opportunities, and community organizations or grassroots movements" (Thomas-Slayter et al. 1996, 295). My own study examines the types of strategies (such as coalition building) employed by leaders in both electoral and grassroots politics. In particular, the literature discussed above guides my research by providing a rich theoretical framework for examining the ways in which race/ethnicity and gender influence the politics of Hispanas and indigenous women active in the environmental justice movement and in environmental policymaking in New Mexico.

New Scholarship on Women's Leadership in Electoral Politics

The 1990s also saw the publication of numerous volumes investigating the characteristics, trajectories, experiences, influence, and status of women in electoral politics. No doubt, the attention given to women during this period by the popular media and academics alike had more than a little to do with the increasing gender gap[39] among American voters, the public's interest in gender issues stemming from televised broadcasts of the Anita Hill / Clarence Thomas hearings, the media's pronouncement of 1992 as the "Year of the Woman," and the numerous actions taken by President Bill Clinton and First Lady Hillary Rodham Clinton on behalf of women. Regardless of the causes, coverage of candidates and of women holding elected and appointed office appeared to flourish over the decade, as did scholarly research on women's political leadership. In particular, political scientists generated numerous studies that analyzed how U.S. political culture and environment, institutional settings, and opportunity structures constrain or contribute to the formation of women's political leadership. The scholarship outlined below emphasizes the importance of women's participation in electoral politics and illustrates the need for women's representation in the public policymaking process.

In a study that focused on women's campaigns for public office during the 1992 election season, Witt, Paget, and Matthews (1994) documented the particular challenges that women face with regard to their (1) gender roles, (2) success at fundraising, (3) coverage by the media, and (4) impact on public policymaking. In comparable studies, Burrell (1994) and Fox (1997) examined the campaign experiences, strategies, fundraising activities, and policy agendas of female and male candidates running for election to the U.S. House of Representatives. All three studies found that, despite the barriers to winning public office, women candidates are having a significant and long-lasting impact on the electoral process, and that gender dynamics are changing the way electoral politics is played (Fox 1997). Furthermore,

> [b]ecause of women's entrance into the political arena, many male politicians are now considering issues and strategy in ways that they never have before. ...[T]his may mean that male politicians are now addressing "women's issues," whereas previously they have ignored them. The presence of female candidates who raise different agendas and concerns has caused some male politicians to reevaluate their own priorities... [And, if] male candidates do alter their agenda, the nature of substantive political representation changes with the inclusion of female politicians (Fox 1997, 186-187).[40]

In other research on women seeking public office in the United States, Susan J. Carroll (1994) revealed both the multiple barriers and the opportunity structures available to women, but also recommended strategies for developing women's leadership and increasing their presence in electoral politics. Moreover, she found that there were significant differences between women's and men's public policy preferences, and that women in office better represented the gendered concerns of female constituents. Carroll's results were echoed in a study that explored the attitudes, behavior, motivations, policy priorities, and legislative products of women state legislators (Thomas 1994). In comparing female lawmakers with their male counterparts along these dimensions, Thomas concluded that:

> ...women legislators embrace priorities dealing with issues of women, children, and the family. Men do not share this priority list. ...[Further,] women legislators display distinctive policy priorities as their leeway to do so expands. ...Evidence suggests that support for female officeholders and their unique perspectives is related to women reaching a critical mass in any given legislature. [Finally,] based on their fulfillment of many individual, group-based, policy, and procedural goals, women have had both a substantial and distinctive impact on the political arena (1994, 7-8).

Such findings have important implications for the practice of a representative democracy and provide empirical support for increasing the number of women in public office.

Both Carroll's[41] and Thomas' findings were corroborated by research conducted by the Center for American Women and Politics (CAWP) at Rutgers University. The CAWP study focused on the policy goals and political strategies of women in the 103rd U.S. Congress and highlighted their legislative achievements (Dodson et al. 1995). The authors found that

> women [are] making a difference by working within their respective chambers to raise new issues within the Congress, to talk about old issues in new and different ways, to shape legislation so that it reflected this expanded range of concerns, and to influence the fate of legislation, in some cases by just making sure it got to the floor and in other cases by playing critical roles in its consideration on the floor (1995, 24-25).

The congresswomen's policy agendas included such diverse issues as insurance coverage of women's mammograms, sexual harassment of women in the military, Individual Retirement Accounts for homemakers, evidentiary rules in the criminal trials of sex offenders, the Family and Medical Leave Act, the Violence Against Women Act, the Freedom of Access to (abortion) Clinic Entrances Bill, the North American Free Trade Agreement, and the Women's Health Equity Act. Dodson and her colleagues concluded that despite their "outsider" status and small numbers, women elected to congress have (1) worked cooperatively to develop women's legislative leadership, (2) placed a broad range of issues of particular interest to women on the public policy agenda, (3) begun to influence how legislative politics is played, and (4) collaborated to successfully pass legislation important to women (1995). Similarly, a recent study of female policymakers in 22 industrialized nations indicates that women in leadership positions around the globe are increasingly exhibiting a commitment to promoting women's policy issues (O'Regan 1997). Consequently, as women enter politics in larger numbers, more policies are being formulated that address women's particular social and economic concerns.

While none of the research cited above considers the race or ethnicity of the office holders (or their identification with their racial/ethnic group), it provides a valuable theoretical foundation for my study of American Indian and Hispanic women leaders. This scholarship informs my investigation of how and why American Indian women and Hispanas enter formal institutional politics, and suggests the types of issues they are likely to advocate and how they might influence public decision-making. Whether the New Mexico leaders hold elected or appointed positions or head grassroots organizations, we can compare and contrast their political trajectories, their policy preferences, their support for women's issues, and their strategies with those of the (presumably white) officials whose opinions, attitudes, and experiences have been reported in the literature. We will revisit the research cited here in later chapters, as I discuss how my findings enlarge upon the existing knowledge and/or challenge what we know about women and politics.

Scholarship on the Politics of American Indian and Hispanic Women

In contrast to the growing body of literature on (primarily white) women in American politics, scholarship addressing the politics of American Indian women and Hispanas in the Southwest remains relatively sparse (McClain and Stewart 1995; Sierra and Sosa-Riddell 1994). Both groups of women, however, have a rich history of political involvement in the life of their communities. In particular, Native American women have struggled for policy reforms to attain tribal sovereignty, cultural preservation, and control over their native lands and natural resources (Sen 1995; Ford 1990; Gunn Allen 1986; Willard 1984; Waldowski 1980-81). Jaimes, whose scholarship provides a great many examples of such women's leadership, states,

> women have always formed the backbone of indigenous nations on this continent. ...[I]t is women who have formed the very core of indigenous resistance to genocide and colonization since the first moment of conflict between Indians and invaders (1992, 311).

While Mexican American women in the Southwest have also rebelled against the changes imposed by the Anglo colonizers, the literature tends to focus on their role as activists in the labor movement organizing for workers' rights, equitable pay, safe working conditions, and fair treatment (Marquez 1995; Fernández-Kelly and García 1990; Rose 1990; Kingsolver 1989; Zavella 1988; Coyle et al. 1980). Both American Indian and Hispanic women activists have traditionally engaged in struggles to remedy social and economic concerns; this work continues through their leadership in contemporary environmental justice organizations.

In seeking to achieve environmental and social justice for their people, indigenous women's activism incorporates such diverse issues as education, crime prevention, health care, racism and civil rights, cultural preservation, environmental protection, economic security, women's and girls' empowerment, and community services (Naples 1998b; Pardo 1998; Hoikkala 1995; Sen 1995; Gunn Allen 1986. Also see Chiste 1994; Jackson 1993; Remal 1991). Whether they are Mexican American mothers organizing against the placement of yet more undesirable land uses in their *barrio* (Pardo 1990) or Indian women fighting the federal government over deployment of nuclear weapons on tribal lands (Redhouse 1984), these women share common approaches to leadership and their activism is often similarly motivated. As Ford states in her study of contemporary Native women activists,

> women of today have propelled themselves into leadership roles, lobbied Congress, organized networks and associations, beat down the doors of bureaucracy, painstakingly snipped or slashed red tape, filed grievances with state and federal agencies, and filed lawsuits to challenge discrimination (1990, 89).

For the most part, indigenous and Hispanic women are devoted to improving the lot of their communities; they seek to empower others and therefore perceive their own role as one of advocacy, and their organizations tend to be structurally and ideologically democratic (see for example Naples 1998b; Pardo 1998; Hoikkala 1995; Sen 1995; Chiste 1994; Hardy-Fanta 1993; Ford 1990; Koester 1988; Hernandez 1980). In other words, they generally value equality, participation in shared decision-making, and consensus.

While Indian women and Hispanas have traditionally occupied informal positions of leadership in their communities, they are now achieving formal leadership in electoral politics. For example, a study of Mexican American women's participation noted sharp increases in the number of Hispanic women elected to public office (MacManus et al. 1986). The researchers posited that the growth in Chicanas' political activism was due to their higher levels of education, greater incomes, proficiency in the English language, and emergence into the middle-class.[42] A scholar examining the presence of Native women on tribal councils similarly found their educational attainment and income to be critical factors in their political success. Miller (1992) explained that state-sponsored education and technical training have provided Native women with expertise that makes them valuable resources to their tribes and allows them to fill important tribal positions. Their administrative expertise, coupled with the centrality of woman's role within the family network and a favorable economic opportunity structure, have enabled women to attain political office within particular tribes (Miller 1992. See also Knack 1995; Lynch 1986).

As tribal leaders, women have continually challenged federal, state, and tribal authorities to formulate and/or reform policy for the benefit of their communities. For example, as Chief of the Seminole people, Alice Brown Davis fought the federal government to maintain Indian control over Indian education and to retain tribal lands (Waldowski 1980-81). More recently, and a "first" in New Mexico, a tribal Governor sought and obtained state status for her tribe in order to establish water quality standards and mitigate pollution of her pueblo's natural resources by off-reservation polluters (Contreras and Shaw 1993). In her study of tribal leaders, McCoy noted that several women entered tribal politics to make reforms, while a majority ran for office in order to "make things better," "find solutions to tribal problems," and work towards tribal self-determination (1992, 62). Specifically, the leaders' most pressing policy concerns were tribal economic development, health care, education, housing, and tribal/federal relations—in that order (1992, 64). Furthermore, the author found that female leaders conceptualized politics as "a public service and obligation of citizenship" (McCoy 1992, 62)—much as Hispana leaders do, as both my findings and other scholarship indicate (see Prindeville and Bretting 1998; Hoikkala 1995; Hardy-Fanta 1993).

Like native women, Hispanas have sought to effect change on behalf of their communities by serving in public office. In separate studies, Cruz Takash (1993) and Hardy-Fanta (1993) examined the leadership of Latinas in different political roles. Their research uncovered patterns in both the activism of Hispana leaders and in their conceptualizations of politics that established the centrality of gender, race/ethnicity, and community. The findings generally revealed that Latinas in leadership roles accessed resources for their constituents/communities, built and expanded social networks, represented feminist or "women's issues" among others, and raised concerns and provided policy perspectives different from the male norm (Cruz Takash 1993; Hardy-Fanta 1993). Furthermore, by identifying common problems and working cooperatively to develop solutions, the Latina leaders ultimately increased participation in politics as they empowered others to play a role in the civic life of their community. Hardy-Fanta's work, in particular, highlights for us the nexus between grassroots politics and participatory democracy. Equally important, Cruz Takash and Hardy-Fanta have effectively challenged long held assumptions about the nature of politics, the division between grassroots politics and electoral politics, and the validity of the discipline's traditionally narrow focus on electoral politics and institutions. Their work on Hispanas has broadened the scope of research to include marginalized groups by providing new bases for theory about participation, public policy, leadership, and representation.

Compiling a history of indigenous women's political participation in New Mexico was a formidable task. No single comprehensive account of their experiences either prior to or subsequent to American colonization exists. Moreover, with few exceptions, what little has been written of native and colonial Spanish/Mexican women's lives seldom mentions women's role in political life. The relative invisibility of indigenous and Hispanic women in the literature also results from the narrow definition of politics employed in mainstream political science, which serves to limit discussions of politics to governmental institutions, voting and elections. Rather than limiting political activity to a narrow set of behaviors performed in the traditionally male-controlled public domain, conceptualizing politics as a function of human interaction and relationships uncovers the rich variety of political involvement engaged in by women and other marginalized groups (Orleck 1997; Tronto 1993; Bystydzienski 1992a; Tolleson Rinehart 1992; Ackelsberg 1988; Ferguson 1987; Githens 1983). As one scholar explains:

> what is defined as "political"—that is, what is publicly relevant—determines what is available for open discussion, the categories in which people come to understand their experience, and the possibilities they see for resistance. ...If what matters to me is considered not to be appropriate to "politics," then I will tend not to participate in (electoral) political activity. Furthermore, in the absence of a community to validate my perceptions, I may well come to see my own concerns as "merely" personal and profess little interest in politics. ...If the activities I undertake in the larger political context are ignored, or their political significance denied, my frustration may well end in resignation and the process of the production of consent (Ackelsberg 1988, 299).

When politics is confined to formal, legally sanctioned behaviors occurring within specific institutional boundaries, we are left with a superficial and incomplete picture of politics and how it subsumes everyday human intercourse.

Feminist scholars in women studies, sociology, political science, and anthropology, interested in the role of women in social movements, have made substantial contributions to the literature. They have documented the history of women's political participation (McGlen and O'Connor 1995; Naples 1992; Tilly and Gurin 1990; West and Blumberg 1990) and the extent and variety of women's contemporary grassroots activism (Orleck 1997; Basu 1995; Ferree and Martin 1995; Bystydzienski 1992a), argued for the recognition of women's organizing as legitimate political activity (Tronto 1993; Tolleson Rinehart 1992; B.A. Carroll 1989; C.J. Carroll 1989b; Ferguson 1987; Githens 1983), and revealed the diversity of women's political styles and strategies (Pesquera and Segura 1993; Ford 1990; Chai and De Cambra 1989; Hamilton 1989; Morgen and Bookman 1988). This research has demonstrated

> unequivocally that the background variable of gender is one of the most, sometimes *the* most, differentiating factor in studies of political behavior. [This work reveals that] women, to a greater *degree* than

> men, and in different *ways*, initiate, pursue and support issues concerning bio-social production and reproduction, that is, those questions having to do with control over and responsibility for, and care of people and other natural resources (Jonasdottir quoted in West and Blumberg 1990, 7-8. Italics in original).

Much of this new work integrates gender, race/ethnicity, class, and the effects of globalization into the study of women's politics. Due largely to this burgeoning feminist scholarship, studies focusing on women's participation in the environmental policy arena are now receiving more attention in the literature (Prindeville and Bretting 1998; Mack-Canty 1997; Miller et al. 1996; Rocheleau et al. 1996; Seager 1996; Pardo 1995 and 1990; Rodriguez-Trias 1994; Pulido 1993; Cable 1992; Di Chiro 1992; Warren 1987).

The amount of work published on women in politics has increased dramatically over the past 25 years thanks largely to the women's movement and the growth in race/ethnic studies and feminist scholarship. While this has resulted in significant contributions to our knowledge about the political participation of racial and ethnic minority women, scholars agree that substantially more research is needed (Liebowitz and Carroll 1996; Baca Zinn and Thornton Dill 1994; Sierra and Sosa-Riddell 1994; Bystydzienski 1992a; Garcia et al. 1991; Albrecht and Brewer 1990; Campos Carr 1988-89; Green 1980; Lurie 1972). A recent publication of the Center for American Women and Politics urged researchers to make the study of women of color in politics a priority:

> Questions about the specific needs and barriers faced by women of color [in the United States] should be integrated into all aspects of research on women and politics. Research is needed which takes the information that we have about "women's" political activities (usually based on studies of predominantly white samples of populations of women) and determines the ways in which experiences of African-American women, Hispanas, Asian-American women, Native American women, and other women of color are similar to and different from those of white women and each other (Liebowitz and Carroll 1996, 3).

Research on the political participation of women of color as community leaders and policymakers contributes to our understanding of their influence as both non-traditional and formal political actors.

My project reveals the important role of racial/ethnic- and gender identity to American Indian and Hispanic women in New Mexico and their contributions to political life by focusing on the experience, political activities, and policy goals of these leaders. Ideally, the implications of this study will contribute to the development of new theoretical constructs and models for participation that incorporate traditionally excluded groups such as American Indian and Hispanic women. I hope that the findings of this study will be used to broaden the scope and depth of the literature on these women as participants in state, local, and tribal politics generally, and the environmental policy arena specifically. As Virginia Sapiro noted in her now classic essay:

> Taking a serious interest in women and politics will...change some of the ways in which political scientists approach and discuss politics. Women's studies research has already pointed out flaws and gaps in accepted political theories and models. It has also revealed many pieces of conventional wisdom to be false, usually by transforming common and sometimes unstated assumptions into questions and hypotheses. Research on women uncovers critical problems in democratic theory and research... (Sapiro 1981, 713).

We now turn to a discussion of the assumptions, approaches, and research methodology employed in this project.

CHAPTER 2
Design of the Research Project

THIS STUDY SEEKS TO ILLUMINATE THE POLITICS OF HISPANA AND indigenous women leaders in the Southwestern United States by focusing on their involvement in the environmental policymaking arena in New Mexico. In this context, power is exercised by individuals and groups successful at setting the public agenda, influencing decision-making, and ensuring the implementation of policy directives. I undertook this particular project for several reasons. First, American Indian women and Hispanas in positions of leadership in the state facilitate the democratic process by empowering others to participate in public life and by representing marginalized groups' interests (Prindeville and Gomez 1999; Bretting and Prindeville 1998; Prindeville and Bretting 1998). These women make important contributions to their communities and effect social change by influencing the distribution of wealth, resources, and influence in society. Second, by studying the politics and policymaking activities of American Indian women and Hispanas in both grassroots organizations and in governmental institutions, we can gain a greater understanding of their contributions and their social impact. Third, Hispanics and American Indians comprise nearly half of New Mexico's population; a substantial portion of the state's polity and a potential political force.[1] Fourth, as feminist scholars West and Blumberg note:

> Political thought and analysis must incorporate not only gender, but also race and class, as well as other categories that continue to divide and separate human beings... Until it does, it must be seen as incomplete and biased. ...When the literature truly reflects the reality of women's lives, history will reveal the astoundingly creative "political"

> lives many women have led in order to survive in a male-constructed and male-dominated world (1990, 11).

Finally, I concur withCohen, Jones, and Tronto when they write,

> We consider it essential to our struggles against dominant power to decenter the experiences, values, and perspectives of dominant groups. In the field of women and politics, this means decentering the focus on elite women, especially white women, which has continued to dominate research. Only by including those women whose lives do not fit neatly into the categories of what has been regarded traditionally as politics...can we perceive the limits and the profoundly deep problems, as well as the opportunities, in American political life (1997, 4).

PROJECT DESIGN AND USE OF THE GROUNDED THEORY APPROACH

As stated earlier, the political science literature neglected, until recently, to examine the political participation of American Indian and Mexican American women (Liebowitz and Carroll 1996; McClain and Stewart 1995; Sierra and Sosa-Riddell 1994; Acosta-Belén 1993; Jaimes 1992; Garcia et al. 1991; Ford 1990; Campos Carr 1988-89; Green 1980). With the exception of radical feminist scholarship, the impact of important variables such as gender, race, and class on women's political participation, for example, were not adequately researched (Cohen et al. 1997; Liebowitz and Carroll 1996; Bennett and Bennett 1986; Millman and Moss Kanter 1975). Indeed, early examinations of race and/or ethnic politics consisted primarily of comparisons between blacks and whites. This often resulted in the overgeneralization of findings from studies of African Americans to other distinct minority groups.

Studies of women political elites usually focused on male-dominated party politics rather than on women-run community based organizations—even though women have much greater opportunities for involvement and face fewer barriers in grassroots politics than in institutionalized party politics. Consequently, scholarship on women and politics was frequently limited by very small samples due to the dearth of women in elected and appointed positions and the fact that researchers seldom combined grassroots activists and public officials in their samples.[2] Moreover, men's political behavior was the standard by which women's was measured (Etter-Lewis 1991; Cook and Fonow 1990; West and Blumberg 1990; Westkott 1990). As Cohen, Jones, and Tronto explain,

> What has made women marginal has been the way that politics has been defined in mainstream political analysis. Key among the factors that have left women out of the political picture has been the excessive reliance on dominant institutions to mark the space of politics, which marginalizes women's ways of looking at and being involved in politics. Another factor has been that material on different women's political action has been organized in texts in a way that treats women of color or working-class women as deviating from a political norm established by elite white women (1997, 5).

As a result, traditional conceptualizations and research approaches generalized from men to women, and from white women's political experiences to those of all women, regardless of their race or ethnicity.

While there is now a growing body of research on racial and ethnic minority women in the United States, this scholarship is quite new and therefore still limited. Without an established tradition of scholarship in this field, I have had to assemble a diverse collection of interdisciplinary research about women's political activism to provide a framework for my theoretical propositions and to guide my investigation. As a result, I have employed various (primarily qualitative) methodological tools, borrowing from the case study approach, ethnographic research, oral history, and grounded theory for designing the study and for collecting and analyzing my data (see for example Berger et al. 1991; Strauss and Corbin 1990). In addition, I found background reading in feminist methodologies and epistemology to be helpful in building a theoretical framework, formulating a research design, and in developing a personal approach to conducting face-to-face interviews (see Chowdhury and Nelson 1994b; Fonow and Cook 1991; Mies 1991; Cook and Fonow 1990; Haggis 1990; McCarl Nielsen 1990; Stanley and Wise 1990; Harding 1987; Githens 1983; Millman and Moss Kanter 1975).

Grounded theory is especially useful for conducting research in new areas of inquiry as it enables the investigator to build theory from the data up when a mature body of scholarship does not yet exist to guide the research project (Miles and Huberman 1994; Tesch 1990). Stanley and Wise explain the difference between traditional quantitative research approaches in social science and those such as grounded theory, which are qualitatively oriented:

> Deductivism treats experience as a 'test' of previously specified theoretical hypotheses; and so within it theory precedes both experience and research, and these latter two are in a sense predicated upon theory. In...contrast, inductivism specifies a model of research in which theory is derived from research experience and is often referred to as 'grounded theory' (1990, 22).

Furthermore, qualitative methods like grounded theory help to expose gaps or areas of weakness in the existing literature and are more useful than quantitative approaches for investigating the hows and whys of human behavior (Miles and Huberman 1994). For example, in survey research (a quantitative research technique) informants play a passive role in the research process, selecting their responses from a limited number of set choices established previously by the investigator. In contrast, qualitative approaches often incorporate open-ended interview techniques that provide participants with an active role in the research process by encouraging them to formulate their own answers, to seek clarification from the investigator, and to pursue themes or a stream of thought that they feel is relevant. In turn, the investigator is free to seek explanation, to ask "why" and "how," to backtrack or to pursue new directions of inquiry with the research participant.

Tesch outlines ten "principles and practices" for conducting qualitative research:

1. Analysis is not the last phase in the research process; it is concurrent with data collection or cyclic...
2. The analysis process is systematic and comprehensive, but not rigid...
3. Attending to data includes a reflective activity that results in a set of analytical notes that guide the process...
4. Data are 'segmented', i.e., divided into relevant and meaningful 'units'...
5. The data segments are categorized according to an organizing system that is predominantly derived from the data themselves...
6. The main intellectual tool is comparison...
7. Categories for sorting segments are tentative and preliminary in the beginning; they remain flexible...
8. Manipulating qualitative data during analysis is an eclectic activity; there is no one 'right' way...
9. The procedures are neither 'scientific' nor 'mechanistic'...
10. The result of the analysis is some type of higher-level synthesis... (Tesch 1990, 95–97. See also Feldman 1995).

Unlike deductive comparative methods traditionally used by political scientists (see for example Mayer 1989; Ragin 1987; Jackman 1985; Eckstein 1975; Lijphart 1975), the inductive strategy of constant comparison employed in grounded theory is the most appropriate for addressing my particular research questions. Throughout the investigative process, the researcher compares and contrasts the data looking for specific patterns of interrelationship among many categories. The patterns discovered assist in the development of theory and provide an understanding of the phenomena under investigation (Strauss and Corbin 1990). This is particularly helpful for researchers working in "uncharted territory"—new fields of inquiry where little if any theory exists. We will return to the discussion of grounded theory shortly.

SELECTION OF THE PROJECT PARTICIPANTS

Using non-random purposive sampling, a total of 50 women active in New Mexico politics were selected for participation in this study. I obtained their names using a reputational "snowball" technique where each woman interviewed was asked the names of other women involved in environmental politics in the state. The bulk of the participants (45) were leaders known for their involvement in environmental policymaking and/or for their roles as advocates for communities of color. The remaining five leaders favored environmental and social justice policies but their primary political involvement was with other issues.[3] The reliability and validity of my sampling strategy was reinforced when the same women were repeatedly identified as policy leaders by different study participants in numerous organizational settings. Generalization to the larger population of women leaders across the United States is not intended, nor would it be appropriate, using this methodology. As Miles and Huberman explain,

> the most useful generalizations from qualitative studies are *analytic*, not 'sample-to-population'. ...The prime concern is with the *conditions* under which the construct or theory operates, not with the generalization of the findings to other settings. ...We are generalizing from one case to the next on the basis of a match to the underlying theory, not to a larger universe. The choice of cases usually is made on *conceptual* grounds, not on representative grounds (Miles and Huberman 1994, 28. Italics in original).

To obtain the sample, I combined three approaches: theory based, comparable case, and opportunistic selection. Using theory based selection, I sought women for inclusion in the study whose experiences appeared to fit those of women leaders in related research, which also provided examples of models developed by other scholars (see for example Kaplan 1997; Cruz Takash 1993; Cantor and Bernay 1992; McCoy 1992). The comparable case approach, a replication strategy, led me to include women whose profiles were similar to the activists and public officials portrayed in other studies to determine whether my findings confirmed or refuted earlier research (for example Hoikkala 1995; Sen 1995; Pardo 1990). Finally, using opportunistic sampling, I sought out atypical and confirming cases as well as disconfirming cases to bolster confidence in my conclusions. This technique has worked in similar studies to reveal important patterns and commonalities among diverse groups of women leaders (see for example Hardy-Fanta 1993; Fowlkes 1992; Ford 1990). Consequently, I was able to follow "new leads; taking advantage of the unexpected" (Miles and Huberman 1994, 28) by including in the study women whose policy agendas did not necessarily incorporate environmental concerns. As it turned out, I found advocates for environmental and social justice policies in electoral politics as well as in a variety of grassroots groups.

Both indigenous and Hispanic women were interviewed for this project, including volunteers and paid staff of grassroots organizations (activists) as well as appointed and elected officials at various levels of government (public officials). Table 1 provides a breakdown of the racial/ethnic identity of the leaders interviewed as well as the position they held. While the indigenous leaders were evenly split between activists and officials, there were two more Hispanic grassroots activists than public officials. Overall though, the study participants were fairly evenly distributed among staff (13) and volunteers (13) of grassroots organizations, and appointed (12) and elected (12) public officials.

Table 1. Racial/Ethnic Identity of Leaders and Position Held

	Grassroots Activists		Public Officials		
Race/Ethnicity	Staff	Volunteer	Appointed	Elected	Total
Indigenous [1]	6	7	7	6	26
Hispanic [2]	7	6	5	6	24
Total	13	13	12	12	50
Percent	26%	26%	24%	24%	100%

[1] This group consists of 25 Native North American women and one Native Hawaiian woman.
[2] This group of Hispanas included two women who self-identified as "Mestizas," acknowledging both their Native American and Spanish heritage.

While few studies combine these types of political actors, it makes sense to do so.[4] As Evans notes,

> [v]oluntary associations offer not only essential training grounds and sources of revitalization for active public life but also the necessary arena in which to reshape our understanding of the meaning of participation and to draw appropriately on the passions rooted in private life (1993, 132. See also Bullard 1990; Taylor 1990; Dabrowski 1983; Moe 1981).

Furthermore, women have historically had many greater opportunities to achieve and practice leadership within community-based organizations than they have had in formal electoral politics.[5] For this reason, many women in elected or appointed office gained their initial political experience working in grassroots groups. Additionally, incorporating both formal/electoral and informal/grassroots leaders in the study provided valuable opportunities for comparison and contrast among women in politics working in a variety of circumstances towards generally similar goals. As Miles and Huberman explain,

> Multiple-case sampling adds *confidence* to findings. By looking at a range of similar and contrasting cases, we can understand a single-case finding, grounding it by specifying *how* and *where* and, if possible, *why* it carries on as it does. We can strengthen the precision, the validity, and the stability of the findings. We are following a *replication* strategy. If a finding holds in one setting and, given its profile, also holds in a comparable setting but does not in a contrasting case, the finding is more robust (1994, 29. Italics in original).

Lastly, as Katz points out, "By searching for data that differ in kind from instances previously recorded, [such] analytic research creates a picture of the scene researched that is strategically biased toward much greater variation than random sampling would reveal" (1988, 137). Therefore, the picture that we gain of women leaders in New Mexico's political scene is rich, detailed, and specific to their circumstances.

American Indian women and women of Hispanic origin, 18 years or older, comprise 233,488 persons or about 22 percent of New Mexico's total adult population.[6] While the precise number of people who hold leadership positions in communities throughout the state is unknown, we can safely assume that women are a minority within this elite group.[7] Subsequently, the small universe from which to draw a sample resulted in my interviewing many, if not most, of the indigenous and Hispanic women leaders in New Mexico politics. Due to their relatively small number, high level of political activity, aggressive coalition-building, and because they constitute a political elite, the women interviewed for this study were frequently acquainted with each other. To protect their identity, and so that no quote is directly attributable to any individual, pseudonyms have been used throughout.[8]

RACIAL/ETHNIC COMPOSITION OF PROJECT PARTICIPANTS

Twenty-five (96 percent) of the 26 indigenous women interviewed were American Indian, registered with one of 17 different nations. As such, they identified as Bad River Chippewa, Blackfoot, Comanche, Diné (Navajo), Kiowa, Western Shoshone, or as Pueblo from Acoma, Cochiti, Isleta, Laguna, Nambe, Pojoaque, Santa Clara, Santo Domingo, Tesuque, or

Zuni. One leader was native Hawaiian. As she was active in the native Hawaiian movement and identified as a member of a colonized indigenous people, I have grouped her with the American Indian women. The 24 Hispanic women variously self-identified as Chicana, Hispanic, Indigenous, Mestiza, Mexican American, and Spanish. I have classified the three women who called themselves either "mestiza" or "indigenous" as Hispanic since they also identified as "Chicana" and/or spoke Spanish at home.

Table 2. Racial/Ethnic Identity of Leaders and Women in New Mexico[1]

Race/Ethnicity	Total Leaders		NM Females	
	No.	%	No.	%
American Indian	25	50	40,757 [2]	7
Hispanic	24	48	192,731 [3]	35
African American	0	0	8,635 [4]	<2
Euro American	0	0	302,507 [5]	55
Other [6]	1	2	1,193 [7]	<1
Total	50	100	545,823	100

[1] Data on women age 18 and over in New Mexico is taken from Table 28 of the U.S. Department of Commerce *1990 Census of Population—General Population Characteristics, New Mexico* 1990 CP-1-33.

[2] The U.S. Census category includes Eskimos and Aleuts.

[3] The U.S. Census category includes all races of hispanic origin, including 1,615 American Indian women of hispanic origin. So as not to count these women twice, they have been omitted from the category of "American Indian" in my table.

[4] The U.S. Census category is "Black not of Hispanic origin."

[5] The U.S. Census category is "White not of Hispanic origin."

[6] One leader is Native Hawaiian. For purposes of this study, I have grouped her with the other indigenous leaders in the various analyses presented throughout this work.

[7] This includes the U.S. Census categories "Asian," "Pacific Islander," and "other race."

Table 2 illustrates the racial and ethnic make up of the leaders in comparison with that of females, 18 years and older, in the New Mexico population. American Indian women account for approximately 7 percent of adult females, Hispanas comprise 35 percent, and "others," including Pacific Islanders, make up less than 1 percent of women in New Mexico. In combination, these three groups amount to a large minority—43 percent of adult females within the state.

The Organizational Context of Project Participants

Over half of the leaders in this study (26 or 52 percent) were volunteers and/or staff in (primarily) environmental justice organizations,[9] while the other 24 (48 percent) held elected or appointed positions in federal, state, local, or tribal government. The grassroots activists included employees and/or volunteers who were, or had been, policymakers such as directors or board members in non-profit community-based organizations working for environmental and social justice.

Table 3. Focus and Scope of Activists' Organizations

	Grassroots Organization's Scope [1]						
Organization's Focus	Local	State	Tribal	Reg'l	Nat'l	Int'l	Total
Indigenous Env'l Rights	1	1	0	2	0	2	6
Environmental Protect.	0	0	0	2	1	0	3
Sustainable Econ. Dev.	2	0	0	0	0	0	2
Political Empowerment	1	0	0	0	1	0	2
Health and Safety	2	3	2	0	0	0	7
Total [2]	6	4	2	4	2	2	20

[1] The scope of the organizations ranged from the local to the regional, national, and international.

[2] There were 20 rather than 26 organizations for two reasons. First, three leaders were "independent activists." Two of these volunteered their technical assistance and expertise to numerous organizations without belonging to any particular group. The third independent activist volunteered for her own and other Indian tribes. Second, three pairs of leaders were active in the same three groups. In two of these cases, one of the women was a volunteer and the other was a staff member. In the third case, both women held leadership positions as paid staff in their organization.

As Table 3 indicates, the grassroots organizations varied in scope as well as focus. Those with a local scope served a particular neighborhood or an entire city. Some groups worked throughout the state of New Mexico while others were limited to serving the members of particular Indian tribes. Some of the organizations had a larger geographic scope that included several states within the Southwestern region of the country. A few of the organizations represented groups throughout the nation, while others reached across national boundaries to serve even broader constituencies. In general, the majority of the activists'organizations (14 or 70 percent) maintained a local, state, or regional scope focusing on indigenous environmental rights and/or issues pertaining to public health and safety (13 or 65 percent).

Although some groups had broader objectives, their organizations generally had a particular focus, which I grouped into five broad categories: (1) Indigenous Peoples' Environmental Rights, (2) Environmental Protection, (3) Sustainable Economic Development, (4) Political

Empowerment, and (5) Community Health and Safety. A focus on indigenous peoples' environmental rights generally included advocating for religious and cultural preservation, repatriating cultural and religious artifacts, returning land, and returning land-, water-, and mineral rights to indigenous tribes. A focus on environmental protection involved conserving and/or protecting valued natural resources and maintaining or improving the quality of the air, water, and/or soil. A focus on sustainable economic development involved creating and operating programs that generated economic opportunities for residents that improved the quality of their lives without harming the environment or the fabric of the community. A focus on political empowerment generally included designing and administering programs that encouraged self-determination, that developed leadership skills and confidence, that taught citizenship, and that increased public participation in decision-making. A focus on community health and safety incorporated concerns as diverse as environmental health, community morbidity, workers' safety, neighborhood safety, and domestic violence. Although I have distinguished the organizations by their principal focus, it is important to realize that many of these grassroots groups had more than one focus.

Table 4. Position Held by Public Officials

Position Held Most Recently	Type of Office	Total
National Advisory Council Member	Appointed	1
District Court Hearing Officer	Appointed	1
District Court Judge [1]	Appointed/Elected	1
Secretary of State	Elected	1
State Board Member [2]	2 Appointed / 1 Elected	3
State Commissioner	Appointed	2
State Department Director	Appointed	2
State Legislator	Elected	3
Tribal Council Official [3]	1 Appointed / 4 Elected	5
City Clerk	Elected	1
City Commissioner	Appointed	1
City Council Member	Elected	1
City Department Director	Appointed	1
Local School Board Member	Elected	1
Total	12 Appointed / 12 Elected	24

[1] This leader was initially appointed to the state District Court by the Governor. She has retained her office by winning subsequent elections.

[2] Some state boards, such as the Board of Education, include both elected and appointed seats.

[3] In order to protect the identity of these leaders, I have not specified their positions within their tribal governments.

Several of the public officials had gained valuable political experience as former community activists. The public officials included elected and appointed officials who had campaigned on issues related to environmental quality, and/or who had served on formal legislative committees, or in federal, state, tribal, or local agencies addressing environmental and social justice concerns. As Table 4 shows, the public officials were a very diverse group comprising elected or appointed leaders at the local, county, state, tribal, and national levels of government. While most of the officials held positions in the legislative or executive branches, two women worked in the judicial branch.

A minority of these leaders (4 or 8 percent) did not directly deal with such policymaking in their official capacity, although they may have done so in previously held positions. As elected or appointed officials, these leaders were generally responsible for representing a broad range of constituent interests. All 24 public officials, however, personally (unofficially) advocated environmental and/or social justice policies.

DATA COLLECTION: THE USE OF PERSONAL INTERVIEWS

The use of loosely structured interviews, administered in person and employing open-ended questions, emphasizes the life experience of project participants allowing them to tell their own stories, to offer their own subjective meanings, and to link these meanings to their understanding of the social world (Feldman 1995; Lofland and Lofland 1995; Miles and Huberman 1994; Anderson and Jack 1991; Hale 1991; Minister 1991). Holstein and Gubrium, who advocate the use of what they call the "active interview," explain that the interview guide

> ...is not so much dictated by a predesigned set of specific questions as it is loosely directed and constrained by the interviewer's topical agenda, objectives, and queries. ...The respondent actively composes meaning by way of situated, assisted inquiry [with the result that]...[t]he respondent becomes a kind of researcher in his or her own right... (1995, 29).

The flexibility offered by employing such open-ended interview strategies enables the field researcher to add, delete, or rephrase questions, to pursue alternate lines of inquiry, to dig deeper, to seek explanation and/or clarification from their source—an important check of validity, and to elicit the participant's own interpretation of the data (Borland 1991; Strauss and Corbin 1990; McCraken 1988; Dexter 1975). The use of open-ended interviews is especially fruitful in conducting research on understudied or marginalized groups such as women of color because, as Kelly, Ronan, and Cawley note,

> In studying women's issues and personal and public experiences [qualitative] analyses provide access to what many scholars assert are fundamentally different gender understandings which remain opaque to standard survey research methodologies and instruments (1987, 25. See also Benmayor 1991; Etter-Lewis 1991).

Anderson and Jack outline some of the benefits of using oral history, which make it especially suited to conducting qualitative feminist research.

> Realizing the possibilities of the oral history interview demands a shift in methodology from information gathering, where the focus is on the right questions, to interaction, where the focus is on the process, on the dynamic unfolding of the subject's viewpoint. It is the interactive nature of the interview that allows us to ask for clarification, to notice what questions the subject formulates about her own life, to go beyond the conventional, expected answers to the woman's personal construction of her own experience. This shift of focus from data gathering to interactive process affects what the researcher regards as valuable information. Those aspects of live interviews unavailable in a written text—the pauses, the laughter—all invite us to explore their meaning for the narrator (1991, 23).

This approach is particularly applicable in feminist research where women's personal narratives can be used to systematically interpret, for example, the political models that women offer (Berger Gluck 1991; Anderson et al. 1990; Buker 1987. Also see Naples 1998a; Orleck 1997; Rocheleau et al. 1996; Mies 1991; Pardo 1990; Bookman and Morgen 1988). Such interview data can reveal the complexities of human experience and emotion providing context, authenticity, specificity, and vivid description.

In summary, the flexibility afforded by combining inductive and deductive methods, as in the grounded theory approach, allows for continual refinement of the interview schedule, clarification of questions and responses (because the researcher continually moves back and forth between data collection and analysis), inclusion of field observations in the analysis, and the exploration of previously unexamined concepts (Holstein and Gubrium 1995; Lofland and Lofland 1995; Miles and Huberman 1994; Strauss and Corbin 1990). In this way, the findings inform theory. As Tesch states, "a single research project does not produce an entire social theory, but it can develop a set of theoretical propositions [or]...statements about relationships" (1990, 85). Accordingly, I employed a combination of inductive and deductive techniques for developing theoretical propositions, data collection and analysis, and theory building.

ENGAGING IN THE INTERVIEW PROCESS

Upon meeting with each participant, the interview process was briefly explained and their consent was obtained in writing. I met with the leaders at various locations throughout New Mexico between April, 1994 and June, 1996, with the exception of three interviews conducted in November, 1991 as part of a pilot project. The interviews lasted between approximately 45 minutes and 2 hours with an average length of one hour. Interviews were tape recorded with the permission of the participant, and hand-written notes were taken simultaneously. A transcribed record of each interview was later made from my field notes and audio tapes. In cases where it was neither possible nor desirable to tape the interview, I relied on my handwritten notes. Participants were offered the opportunity to review their transcribed interview. In two cases, brief follow-up calls or meetings were held to complete or clarify parts of an interview. Some leaders preferred to meet in their home, but most were interviewed at either their workplace or at a public spot such as a coffee shop.

Two interview guides were prepared, one for the activists and one for the public officials with only minimal variations to account for the leaders' different circumstances. (See Appendices A and B.) While the interview guides followed a similar set of questions, participants spoke in their own words and were encouraged to pursue related issues expounding on areas they thought relevant. As Holstein and Gubrium advise:

> A rule of thumb for using an interview guide is to let the respondent's responses determine whether particular questions are necessary or appropriate as leading frames of reference for the interview conversation. This lends a rather improvisational, yet focused, quality to the interview... (1995, 77).

Both the format of the guide and the interview process stressed participants' definitions of concepts and issues, encouraged them to structure their own account of the problem or policy concern, allowed for detailed descriptions by the participants, and facilitated their communicating their particular idea of what is relevant (Lofland and Lofland 1995). Finally, the flexible interview format gave participants special, non-standardized treatment consistent with the research goal of developing as rich and as accurate a profile as possible of each of the leaders (Feldman 1995; McCraken 1988; Dexter 1975).

In particular, the interview guides were intended to accomplish numerous objectives. First, in order to develop a biographical sketch of each participant, information was collected about their demographic characteristics, the history of their political involvement, and their current life situation. Holstein and Gubrium explain the rationale for this strategy:

> Background knowledge of circumstances relevant to the research topic and/or the respondent's experience can be an invaluable resource for the interviewer. ...Background knowledge allows the interviewer to move from the hypothetical or abstract to the very concrete by asking questions about relevant aspects of the respondents' lives and experience, a particularly fruitful tactic for promoting circumstantially rich descriptions, accounts, and explanations. The general point is to engage respondents in meaningful talk about their everyday worlds in terms that derive from the circumstances of lived experience (1995, 77).

Second, questions were designed to explore the leaders' political socialization, their experiences of "politicization,"[10] their ideological orientations, and their motives for participation. Third, the interview process sought to identify each leader's political goals and public policy agenda, their roles in the policymaking process, and whether their racial-, ethnic-, and/or gender identity influenced their policy objectives. Lastly, the interviews identified leaders' diverse strategies for social change and their perceived effectiveness in accomplishing their goals.

GENDER, RACE, ETHNICITY, AND CULTURAL CONSIDERATIONS

The researcher must be sensitive to gendered and cultural communication differences. Besides the obvious potential for miscommunication arising from the interviewer and interviewee's proficiency in different languages, the researcher must be alert to subtle yet critical differences in their use of, and the meanings attributed to, key concepts. For example, I found that participants did not necessarily define or understand ideas central to my project such as "politics," "feminism," "activism," and "environmentalism," in the same way that I did. Therefore, obtaining valid and reliable responses required a certain amount of follow up questioning, explanation, and asking for elaboration of critical terms. The potential for interviewer bias is exemplified in the following analysis.

> As a researcher, I have learned that critical areas demanding attention are frequently those where I think I already know what the [respondent] is saying. This means I am already appropriating what she says to an existing schema, and therefore I am no longer really listening to *her*. Rather, I am listening to how what she says fits into what I think I already know. So I try to be very careful to ask each [respondent]

> what she means by a certain word, or to make sure that I attend to what is missing, what literary critics call the "presence of the absence" in women's texts... (Anderson and Jack 1991, 19. Italics in original).

The importance of recognizing the "absence" in women's narratives is illustrated in the following story. An Indian activist arrived with her husband for an interview with me at a local coffee shop. As the interview progressed, I became increasingly frustrated by her seeming lack of cooperation and unwillingness to expound on a series of short, blunt answers to my open-ended questions. I began to wonder whether I had made a mistake by including this woman in my sample of leaders as she continually downplayed the importance of her role in the native rights movement. Her husband, however, was cognizant of the cultural differences being played out in the dynamics of our exchange, and fortunately, he intervened.

The leader's husband pointed out many of his wife's substantial contributions to and accomplishments in the native rights struggle, successfully eliciting explanations and rich descriptions from her about her political activism, and her feelings about these experiences. Thinking about our meeting later on, I became acutely aware of the uncomfortable position in which I had placed the leader; in essence, I was focusing the conversation on her and asking her to "boast" about the significant parts she had played during her 26 years as an advocate for American Indian rights. I had erroneously interpreted her embarrassment at being placed in this position as reluctance to cooperate, when in fact, her discomfort was more likely a function of my asking her to violate the cultural and gendered norms that establish "acceptable" forms of women's communication. Had her husband not encouraged her revelations, I would have left the interview with a highly inaccurate picture of the extent of this leader's political participation.

The above experience demonstrates how women of color might exhibit styles of communication distinct from those of the dominant Euro American culture and/or different from those of men (Minister 1991). Consequently, the ways in which a respondent's communication style is shaped by her culture and gender should be considered both prior to and during question design, data collection, and analysis of the findings. Similarly, the sex, race, and ethnicity of the interviewer may have an effect on the outcome of the interview. For example, women are likely to share at least some significant gendered experiences such as sexual discrimination, menses, childbirth, menopause, battery, or rape (Minister 1991; Stanley and Wise 1990). The fact that I was visibly pregnant during many of the interviews and that I later mentioned to most, if not all, of the participants that I was a new mother, likely helped to establish a common bond between us. Holstein and Gubrium explain how to build on such commonalities by strategically joining in an exchange with project participants.

> Conversational give-and-take around topics of mutual interest is a way of conveying to the respondent that the interviewer is sensitive to, and interested in, the ongoing line of talk. Drawing on mutually familiar events, experiences, or outlooks not only secures rapport...but fixes the conversation on particular horizons of meaning or narrative connections, encouraging the respondent to elaborate (1995, 77).

At the same time, as one researcher cautions and my own experience revealed, it is important not to make too much of assumed similarities between the interviewer and interviewee.

> [R]egardless of shared ethnicity, gender, or class, a researcher enters each and every field situation as an "outsider." We [must be] acutely aware of points of convergence and unbreachable gaps between ourselves and the people with whom we are working (Benmayor 1991, 171).

Along similar lines, while most of the women I interviewed were politically sophisticated, highly polished, and used to speaking to the media and to groups on a regular basis, other participants were not initially at ease with me. I perceived that in many instances, women in this latter group adopted the part of "teacher," "elder," "aunt," or "storyteller" during their conversations with me. In response, I assumed the role of "student," "junior," "niece," and active listener. This was always a conscious decision, as I sought ways to bridge the distance and power imbalance between us. As one scholar notes:

> ...Field-workers can analyze the hierarchical system brought to specific interview situations, asking themselves, "How can I equalize the power inherent in the differences between my narrator's and my age, class, ethnic affiliation, and education?" (Minister 1991, 35)

I observed that the interviews conducted in women's homes, or in public places such as coffee shops, were nearly always longer, and often afforded a greater degree of intimacy and comfort for the respondent. In contrast, meetings held in the respondent's workplace were more likely to remain formal. It seemed that as long as a leader maintained her official persona, rather than revealing herself primarily as, for example, a woman, Chicana, mother, or wife, a perceivable distance remained between us. This was not always the case, however, and whenever possible, I tried to establish a connection between us.[11]

Our shared gendered experiences may have contributed to creating an environment more conducive to intimacy and self-revelation. Indeed, numerous leaders wept in my presence when relating painful memories about their struggles in politics. Others shared intimacies about their family life, partner, or financial circumstances. Three participants revealed to me their pasts as battered women; another spoke with great emotion about an aborted pregnancy. Many women were warm and friendly, cracking jokes and letting me into their lives through their narratives, for which I am extremely grateful. I was both awed and honored by their trust in me and found myself wanting to give something back to the women; much as Acker, Barry, and Esseveld relate from their own experience:

> ...Part of the attempt to deal with the subject-object problem was to try to establish some reciprocity by offering, at the end of the first interview, to tell the women something about ourselves if we had not done so earlier. Often we didn't have to offer—it was a request made of us. ...We formed friendships with many of the women in the study. We were offered hospitality and were asked to meet husbands, friends, and children. Sometimes we would provide help to one or another of the women in the study (1991, 141).

In an effort to equalize the relationship between researcher and respondent, I volunteered to share information about my project and myself following our interview, if the participants so desired. Furthermore, as a scholar engaged in activist research, following their interviews, I offered respondents copies of my published findings, assistance with political campaigns, and support in demonstrations or with other forms of direct action. I also invited particular leaders to speak to my classes and even created an opportunity for some of the women to network at a casual gathering held in my home. However, I received far more from the project participants than I could ever give; they generously gave up their time, welcomed me into their homes and workplaces, and graciously answered my many questions.

While qualitative forms of inquiry have the potential to reduce participants to objects of study as easily as traditional quantitative methodologies, the researcher's approach and attitude can make a big difference to the participants' feeling either exploited or respected and valued. Some feminist theorists go so far as to suggest that participants take an active role as "co-researchers" and/or co-authors of their research studies (Acker et al. 1991; Borland 1991). Others assert the importance of maintaining distance between informant and researcher to ensure the integrity of both the participant and the study (see for example Hale 1991; Patai 1991). For example, Acker, Barry, and Esseveld advise:

> ...The active voice of the subject should be heard in the account. Our interpretation should avoid transforming the acting and thinking human being solely into an object of study, while recognizing that some objectification is inherent in the process of interpretation or reconstruction [of the data] (1991, 145).

The nature of qualitative techniques provides opportunities for both participants and researchers to take an active role in this dynamic process, acknowledging the contributions of participants, minimizing the chances of their feeling exploited, and respecting their dignity as individuals. Many of the women participating in my project shared volumes of information and all provided valuable insights revealing their tremendous knowledge and experience in grassroots and electoral politics. As an activist scholar, my commitment to them is to tell their stories, to gain them recognition for their substantial and vital work, and to provide young women with role models of effective, powerful, and caring political leaders.[12]

We now turn to a discussion of the environmental justice movement in New Mexico, its history and goals, to provide a context for examining indigenous women's and Hispanas' leadership in the environmental politics of the state.

Chapter 3

New Mexico's Environmental Justice Movement

The environmental movement in the United States was originally formed during the late nineteenth century. Affluent Anglo Americans were anxious to protect wilderness areas for aesthetic and recreational purposes from the detrimental impact of urbanization and the industrial revolution.[1] The modern environmental movement grew out of a resurgence in ecological concern and the social activism of the 1960s. Its leaders too were largely Euro Americans concerned with aesthetics and with attaining sustainable growth while preserving America's natural resources (Caulfield 1989; Faich and Gale 1971). The goal of managed economic expansion through conservation appealed to conservative and liberal environmentalists alike and contributed greatly to the success of the movement. Less obvious was the distributional effect or socio-economic impact of conservation policies. As early as 1970, at Harvard University's Earth Day rally, a civil rights leader and welfare activist predicted the chasm between the elite white membership of the environmental movement and the poor and people of color:

> I'd like to share with you some concerns that I have about the movement around the environment... If you are a serious movement, you must be prepared to take on the giant corporations who are the primary polluters and perpetrators of some of the worst conditions that affect the environment... [But] it has been my experience that most of you aren't going to deal with the problem at the level of the welfare recipient, the poor person in the ghettos and the barrios. Most of you are not even going to listen to the voices coming from those communities. You won't ask them what they want, and how they want to deal with the problems of their environment, or indeed whether they want

> to deal with the problems of environment at all—because they feel there are more pressing priorities in their lives. Is the ecology movement planning to place any serious priority on the problems of environment in the ghetto and the barrio, of our urban areas, where pollution is worse? (George Wiley quoted in Citizen's Clearinghouse for Hazardous Waste, Inc. 1990, 8).

In fact, the uneven enforcement of existing environmental and land use regulations, the lack of support on the part of environmentalists for the plight of communities of color, and the failure of environmental organizations to recognize local expertise have all contributed to the perception that the environmental movement is elitist and racist (Lawson 1995; Salazar and Moulds 1995; SouthWest Organizing Project 1995; Gedicks 1993; Hamilton 1993; Peña 1992; Bryant 1991; Head and Guerrero 1991; Bullard 1990; Carter 1990; Taylor 1990; Bullard and Wright 1987). The mainstream movement has also been faulted for being sexist. As one leader[2] in New Mexico's environmental justice movement noted,

> there is some [sexism] in the environmental community. It's like a microcosm of American society. There are mainly male leaders, they leave out women-led groups and women who'd want to get involved. It's American culture: they're not used to women taking leadership roles.

In general, leaders of the mainstream environmental groups believed that racial/ethnic minorities had little interest in their organizations' environmental agendas. And, to some extent, academic scholarship in the late seventies and early eighties contributed to this misconception with a plethora of conflicting findings among studies, few of which actually controlled for the race, ethnicity, or sex of respondents. Much of this research either addressed the salience of environmental issues to individuals with differing socioeconomic status and/or political ideology, or examined demographic factors leading to public support for environmental protection (see for example Hamilton 1985; Cotgrove and Duff 1980). Those early studies that did include such variables produced mixed findings about the effects of race, income, education, age, occupation, residence, and sex on individuals' environmental beliefs (Lowe and Pinhey 1982; Van Liere and Dunlap 1980). For example, where some findings indicated that African Americans are less concerned than whites about pollution (Lowe et al. 1980; Hershey and Hill 1977-78), other research found high levels of concern among racial/ethnic minorities and people with low incomes (Adeola 1994; Taylor 1990 and 1989; Wilson 1989; Cutter 1981; Mitchell 1979). Similarly, where more educated, young, urban people exhibited greater awareness of environmental problems and support for reforms

(Neiman and Loveridge 1981; Van Liere and Dunlap 1980; Buttel and Flinn 1978), other researchers found "no evidence that younger, high income, or highly educated residents are more involved in environmental practices" (Baldassare and Katz 1992, 612; also Cutter 1981). Finally, where some findings revealed that women express higher levels of concern over environmental threats (Hamilton 1985), or are more likely to engage in environmental practices (Baldassare and Katz 1992), other evidence indicated that sex has a negligible influence on environmental concern (Adeola 1994; Van Liere and Dunlap 1980).

Since the late eighties, a growing body of scholarship has revealed the importance of examining race/ethnicity and exposure to environmental threats (Allen et al. 1995; Cable and Cable 1995). It also documents the origin of the grassroots environmental justice movement in Southern black communities and its roots in traditional civil rights agendas (see for example Salazar and Moulds 1995; Cable and Benson 1993; Freudenberg and Steinsapir 1991; Alston 1990; Bryant 1989; Hall 1988; Bullard and Wright 1987). Furthermore, it provides considerable evidence of the disproportionate risk of racial/ethnic minorities and the poor to environmental hazards (Tesh and Williams 1996; Westra and Wenz 1995; Adeola 1994; Hutchings 1994; Rodriguez-Trias 1994; Bullard 1993; Paehlke and Rosenau 1993; United States Environmental Protection Agency 1992; Bryant and Mohai 1990). Disproportionate risk, also referred to as "environmental racism," is best described as

> racial discrimination in environmental policymaking...in the enforcement of regulations and laws...in the deliberate targeting of communities of color for toxic waste disposal and the siting of polluting industries...in the official sanctioning of the life-threatening presence of poisons and pollutants in communities of color...[and] in the history of excluding people of color from the mainstream environmental groups, decisionmaking boards, commissions, and regulatory bodies (Chavis 1993, 3).

Most importantly, this literature identifies the relationships between race/ethnicity, socioeconomic status, political empowerment, environmental concerns, exposure to environmental hazards, and involvement in social movements (Allen et al. 1995; Cable and Cable 1995; Gedicks 1993; Paehlke and Rosenau 1993).

THE EMERGENCE OF THE ENVIRONMENTAL JUSTICE MOVEMENT

During the 1970s, the environmental movement succeeded in gaining passage of 23 different environmental laws at the federal level alone. In response to these new, stricter health and environmental regulations, industries began claiming that environmental goals could be reached only be reducing wages, cutting jobs, and closing plants. A depressed economy also contributed to the unions' weakened bargaining position and their reluctance to press for improved safety conditions. The resulting conflict of jobs versus environment has been referred to by some scholars as "job blackmail" (Bullard 1990, 61). Taking advantage of this situation, industries began offering poor communities compensation and monetary inducements for siting hazardous, toxic, and municipal waste facilities in their jurisdictions (Ambler 1991; Head and Guerrero 1991; Bullard 1990; Heiman 1990). For example, in their examination of toxic wastes and race, the United Church of Christ Commission for Racial Justice reported that:

> [t]he hazardous waste issue, as well as many other environmental problems, has become very much linked to the state of the economy in a given community. These communities have been and continue to be beset by poverty, unemployment and problems related to poor housing, education, and health. ...Consideration of the racial and socio-economic status of a community when dealing with hazardous wastes is critical from a public health perspective (Lee 1990, 13).

Threats of "job blackmail," made by government as well as industry, resulted in disadvantaged communities accepting risks that more affluent communities could avoid.

In 1983, the U.S. General Accounting Office (GAO) first reported a significant relationship between race and the siting of hazardous waste disposal facilities following a survey of commercial hazardous waste landfills located in eight Southeastern states (Lee 1993; Alston 1990). In particular, the GAO noted that although African Americans constituted only 20 percent of the region's population, the majority of the residents living near three of the four commercial landfills were black. What's more, approximately one third or more of the population of these communities lived below the poverty level. At the same time, residents across the country began reassessing the benefits of economic compensation against the health and social costs incurred from hosting landfills, incinerators, and other noxious facilities (Alston 1990; Bullard 1990).[3] Many people felt trapped in polluted environments because of low paying jobs, residential segregation, and discriminatory zoning and land use policies (Lawson 1995; Human Environment Center 1991). Increasingly aware of the impact and stigma of these sites and of the disproportionate risk that communities of color face, grassroots groups began organizing around issues of social and economic equity and environmental justice (Tesh and Williams 1996; Cable and Benson 1993; Ambler 1991; Anaya and Chavis 1991; Moreno 1991; Schwab 1991; The Panos Institute 1990; Heiman 1990).

In 1987, four years after publication of the GAO's survey, the Commission for Racial Justice of the United Church of Christ (UCC) released their report, "Toxic Wastes and Race: A National Report on the Racial and Socioeconomic Characteristics of Communities with Hazardous Wastes." First, the findings documented a striking relationship between race/ethnicity and the siting of commercial hazardous waste facilities. The single variable that best explained the presence or absence of these facilities in a community was the racial composition of community residents. Second, the report demonstrated the prevalence of unmonitored toxic waste sites in racial/ethnic minority communities across the United States. Specifically, the UCC document indicated that over 15 million African Americans and over 8 million Hispanics, or three out of every five blacks and Hispanics, lived in communities with one or more unmonitored toxic waste sites (Lee 1993, 49. See also Cable and Cable 1995; Bullard 1993; Lee 1993 and 1990; Alston 1990; Bryant and Mohai 1990; Bush 1988; United Church of Christ Commission for Racial Justice 1987). The study concluded by calling on all levels of government to make the elimination of hazardous wastes in African American, Hispanic, and other communities of color a policy priority.

The UCC landmark study is responsible for codifying the term "environmental justice," and continues to influence the expansion of this social movement. Environmental justice refers to

> the belief that both environmental benefits and...costs should be equally distributed in society, and that corporations should be obligated to obey existing laws, just as individuals are so obligated (Cable and Cable 1995, 107).

Environmental justice groups seek to ensure that locally undesirable land uses (LULUs) are not disproportionately located in poor and minority neighborhoods and that these communities are not targeted for placement of these facilities (Tesh and Williams 1996; Bullard 1995 and 1993; Cable and Benson 1993; Moreno 1991; Heiman 1990; Bryant 1989; Hall 1988). Emphasizing the fair distribution of the costs of environmental impacts, the movement:

> (1) ...focuses on equity, (2) it challenges mainstream environmentalism for its tactics but not its goals, (3) it emphasizes the needs of community and workplace as primary agenda items, (4) it uses its own self-taught "experts" and citizen lawsuits instead of relying on legislation and lobbying, (5) it takes a "populist" stance on environmental issues relying on active members rather than dues-payers from mailing lists, and (6) it embraces a democratic ideology akin to the civil rights and

> women's movements of the sixties (Gottlieb and Ingram 1988, 14. See also Hall 1988; Bullard and Wright 1987; Gale 1983).

Supporters of the environmental justice movement argue that poverty and the inequitable distribution of resources pose tremendous threats not only to public health and the environment but also undermine the stability of social and political systems.

One scholar/activist has suggested that governments should espouse the following five precepts of environmental justice in order to ensure equal environmental protection:

> guaranteeing the right to environmental protection, preventing harm before it occurs, shifting the burden of proof to the polluters, obviating proof of intent to discriminate, and redressing existing inequities (Bullard 1995, 9).

While these particular principles have not been universally adopted by government, the impact of the environmental justice movement has been felt nationally (Allen et al. 1995). As Costain notes, social movements, such as the environmental justice movement, play a critical role in democratic politics because they

> raise serious questions outside normal government channels, often concerning subjects which are not raised as topics of political concern. The response of the government to demands by movements often results in expanding the areas of government interest and involvement (1981, 100).

This, in fact, occurred with regards to the environmental justice movement and the federal government. By 1992, the head of the U.S. Environmental Protection Agency (EPA), William Riley, was using the term "environmental racism" and the EPA's official publication was addressing the issue of equal environmental protection for the nation's racial/ethnic minorities (See United States Environmental Protection Agency 1992). Later that same year, the Office of Environmental Equity was established within the EPA. And, upon assuming office in 1993, the EPA's administrator, Carol Browner, made environmental justice a priority and set the following goals for the agency:

> 1. Environmental justice must be integrated fully and consistently into the Agency's policies, programs and activities.
> 2. Additional research is needed to address human health and environmental risk to high-risk populations, including the identification of multiple and cumulative exposures or synergistic effects.

3. Environmental data must be collected, analyzed and disseminated routinely. This is particularly true for data comparing environmental and human health risks to populations identified by national origin, income and race.
4. Compliance monitoring, inspections and enforcement actions must be targeted and have a multi-media focus.
5. There must be early involvement in the Agency's activities by all stakeholders and information on human health and the environment should be clear and readily accessible to all stakeholders. (Gaylord and Bell 1995, 36).

Moreover, in 1994, President Bill Clinton issued an executive order entitled "Federal Actions to Address Environmental Justice in Minority Populations and Low-Income Populations." It required all federal agencies to

> achieve the principle of environmental justice by addressing and ameliorating the human health and environmental effects of the agency's programs, policies and activities on minority and low-income populations in the United States (Allen et al. 1995, 1).

Although the order reinforces laws already in place to prohibit discriminatory practices in programs receiving federal financial assistance, the President's actions demonstrate the considerable influence that the environmental justice movement has exerted on national public policymaking.

THE ENVIRONMENTAL JUSTICE MOVEMENT IN NEW MEXICO

It is estimated that between 6,000 and 10,000 local environmental justice groups exist in the United States (Goldman 1991; Edelstein 1988; Freudenberg 1984). This grassroots mobilization has grown in response to local environmental threats and to the marginalization of working-class, third world community interests by the largely white, middle-class environmental movement (Cable and Cable 1995; Citizen's Clearinghouse for Hazardous Wastes, Inc. 1992; Anaya and Chavis 1991; Alston 1990). These new grassroots organizations, which are smaller in size than national organizations, tend to represent people of color, the working class, and the poor (Cable and Benson 1993; Taylor 1993; Di Chiro 1992; Moreno 1991; Hall 1988).

The environmental justice movement is active in New Mexico in local, state, regional, and tribal politics. The state's history, political culture, its large racial/ethnic minority population, political economy, and location have all contributed to the founding and growth of environmental justice groups throughout New Mexico. The legacy of colonization, first by Spain and then by the United States, has left an indelible mark on New Mexico's culture, economy, and politics. Unresolved conflicts involving Indian sovereignty, Spanish land grants, the U.S. government, and private developers continue to fuel debates over rights to land and water (see especially Rodriguez 1987). Tensions persist between the economic draw of tourism on the one hand, and the commercialization of indigenous peoples and their cultures on the other. What's more, New Mexico's economic dependence on both the federal government and the defense industry has left a legacy of environmental destruction. Specifically, hazardous and nuclear wastes produced through the mining of uranium and the manufacture and testing of nuclear weapons in the state have led to countless deaths due to radiation exposure, and have resulted in significant environmental contamination of soil and groundwater (Fleck 1995; Guerrero 1994; Jones 1994; Seidman 1994; Contreras 1993; Partners in the Environment 1993; Kimball 1991; Taliman 1991b). In addition, the Waste Isolation Pilot Project (WIPP), a national storage site for medium and low level radioactive wastes, has been constructed in New Mexico, despite considerable protest. The combination of these factors and their social, economic, and political impact serve to make the environmental justice movement's agenda especially salient to New Mexicans. Consequently, grassroots organizations have formed throughout the state in both urban centers and rural areas, on and off of Indian lands.

Like similar groups around the country, New Mexico's environmental justice organizations identify concerns arising from environmental and economic conditions such as community development, neighborhood safety, pollution, the availability of low-income housing and public transportation, employment opportunities, and workplace hazards (Salazar and Moulds 1995; Marquez 1994; Head 1993; Human Environment Center 1991; New York Times News Service 1991; Bullard 1990; Lee 1990; Kutz 1989). Local groups have campaigned to influence public policies involving the siting of undesirable land uses (Fleck 1995; Partners in the Environment 1993; Conrad 1991; Taliman 1991a), groundwater contamination (Contreras and Shaw 1993; Anaya and Chavis 1991), industrial air emissions and effluent discharges (Enslin 1991; Moreno 1991), airport noise, soil erosion and toxicity (Taliman 1991b), waste reduction and incineration (Inter-Hemispheric Education Resource Center 1994; Fernandez 1993). Additionally, these organizations are involved in issues surrounding neighborhood and cultural preservation (Naranjo 1994), Indian sovereignty (Jones 1994; Gedicks 1993; SouthWest Organizing Project 1991), indigenous people's rights (Pulido 1993; Peña 1992; Taliman 1991c), and civil rights abuses. These grassroots organizations form coalitions with other groups in a variety of activities: to protect sacred Indian lands (Laws 1995; Taylor 1988), to protest anti-immigration or English-only laws, to boycott union busting companies, or to advocate for community health and safety (Seidman 1994; Contreras 1993; House 1993; Moreno 1991). Besides responding to problems, these organizations work proactively to highlight citizen concerns and to develop solutions (Romero 1997; SouthWest Organizing Project 1995 and 1994a; Rhee 1993; Head and Guerrero 1991).

First and foremost, environmental justice groups deal with issues affecting the quality of people's lives. Initially, their campaigns tend to be issue-specific and highly visible using unconventional means of political expression such as petitions, pickets, demonstrations, and obstructions. Their protests are not designed to negotiate. Although they may start out with the intent of fighting a specific issue, many groups later broaden their scope to address other problems that impact their constituency (Taylor 1990; Bryant 1989; Bullard and Wright 1987). For example, the Sawmill Advisory Board, an Albuquerque neighborhood association, organized in response to significant air and noise pollution generated by a sawmill in the *barrio*. After residents spent considerable time and energy in negotiations with the city and plant owners, they succeeded in shutting down operations until the sawmill complied with municipal health and nuisance ordinances. Since that victory, the association has also pressed the city into paving streets and putting sidewalks in the residential areas of the neighborhood. Furthermore, the board distributes a barrio-wide newsletter, conducts a summer Spanish culture program for school age children, participates aggressively in the city land use planning and economic development processes, and has established strong neighborhood leadership to represent the community (Juarez 1991).

In general, New Mexico's grassroots environmental groups: (1) endorse policies that favor the disenfranchised, (2) focus on equality and on distributional impacts, (3) advocate direct action, and (4) solicit the support of local civic and religious groups (Bretting and Prindeville 1998; Mack-Canty 1997; SouthWest Organizing Project 1994b; Bullard 1993; Moreno 1991). While they vary with regard to membership, objectives, and strategies, environmental justice groups in New Mexico share broad goals and policy agendas with similar organizations around the country (See Salazar and Moulds 1995; Alston 1990; Guerrero and Head 1990; Lee 1990). In fact, numerous New Mexico groups maintain ties with organizations across the United States and in other countries that also seek environmental, economic, and social justice for indigenous peoples. For instance, the Southwest Network for Environmental and Economic Justice (SNEEJ) was established in 1990 as a regional multi-issue coalition by the SouthWest Organizing Project (SWOP), a vocal social justice group based in Albuquerque.[4] The following "Statement of Solidarity" was presented by SNEEJ leaders at the People of Color Regional Activist Dialogue held in Albuquerque the same year. It outlines both the group's mission and the precepts underlying the environmental justice movement.

> We are a multi-cultural, multi-national, grassroots network whose focus is to address the fact that communities of color, as well as economically oppressed communities, suffer disproportionately from toxic contamination. We are deliberately targeted through genocide of indigenous people, the threatening of future generations, racism, sexism, and a lack of economic and social justice (Southwest Network for Environmental and Economic Justice, April 8, 1990).

SNEEJ, described as "an organization of organizations" by a staff member, currently serves about 70 affiliates in the Southwestern United States and Northern Mexico. Together with SWOP, it has been instrumental in (1) convening the First National People of Color Environmental Leadership Summit in 1991, (2) pressing the Clinton Administration to establish the Federal Advisory Committee on Environmental Justice in 1993, and (3) opening a Washington, D.C. office for environmental justice in 1994 (SouthWest Organizing Project 1995. See also Gaylord and Bell 1995; Bullard 1993; Lee 1993).

The Leadership Summit itself was significant in that it brought together people of color from around the United States as well as from Canada, Central and South America, and the Marshall Islands "to develop a process for framing the contours of a 'multiracial movement for change' founded on the political ideology of working from the grass roots" (Di Chiro 1992, 104). The process involved the creation of 17 "Principles of Environmental Justice" including "securing our political, economic and cultural liberation that has been denied for over 500 years of colonization and oppression, resulting in the poisoning of our communities and land and the genocide of our peoples" (quoted by Di Chiro 1992, 105). Women participating in the conference reported that "the most promising achievement of the Leadership Summit was...its commitment to the construction of diverse, egalitarian, and nonhierarchical leadership and organizational processes and structures" (Di Chiro 1992, 105). Furthermore, attendees of the First National People of Color Environmental Leadership Summit produced specific policy recommendations, which were presented the next year at the 1992 United Nations Conference on Environment and Development in Brazil.

For the most part, New Mexico's numerous grassroots environmental organizations are neither as radical as Earth First! nor as tied to the establishment as the Sierra Club (Marquez 1994). Instead, they tend to reflect local racial and class diversity, offering opportunities for political involvement to people of color and the poor, which these other associations do not

(Prindeville and Bretting 1998; Cable and Cable 1995; Taylor 1993; Di Chiro 1992; Hall 1988). Unlike the national organizations, grassroots environmental groups are often formed by women with no previous interest in conventional environmental issues and with little or no prior experience in community organizing. Their coalitions are often made up of working-class and third world people, few of whom have had a history of participating in civic affairs. These individuals become involved and remain active, however, because of the salience of the problems being addressed: jobs, health, family and neighborhood safety.

GRASSROOTS POLITICS: FERTILE GROUND FOR WOMEN'S LEADERSHIP

Historically, social movements have been a primary vehicle for women's political participation (McGlen and O'Connor 1995; Bystydzienski 1992a; Hall 1990; Hamilton 1989; Steurnagel 1987; Flammang 1984b). As one scholar notes, "In the procedural or pluralistic democracy of the United States, organizations and organized interest groups represent the major vehicle for input into decision making regarding policy and resource allocations at all levels of government" (Aragon de Valdez 1980, 7). Successful citizen organizations enable individuals to effect change through collective action. In doing so, the power balance shifts from the impersonal, unilateral decision-making of politicians, business, and technicians to mediated resolutions arrived at through negotiation with community leaders.

This type of grassroots involvement is particularly accessible to women who are already a part of community networks that provide resources, support, and communication links with other residents. These networks are familiar and informal. They build on existing relationships established among women through their neighborhood, children's schools, place of worship, club or civic organization, and/or place of work (Morgen and Bookman 1988; Boneparth 1984; Dabrowski 1983; Tilly and Gurin 1990). Further, community organizing activities employ many of the skills that women develop through their roles as homemaker and mother.

> The family is what social justice movements are all about. ... [Women's] work in managing the home, balancing each family member's needs and interests, dealing with money and all the other tasks builds skills that are often easily adapted and applied to the leadership of a community organization (Gibbs, no date (b)).

Women's involvement in community-based groups also provides occasions for social interaction, status and recognition, an increased sense of personal and political efficacy, confidence-building, and greater personal satisfaction (Hardy-Fanta 1993; Cable 1992; Albrecht and Brewer 1990; Carroll 1989; Fahey 1978; Pateman 1970). As one feminist scholar notes,

> ...women's community-based political activism is a conscious and collective way of expressing and acting on their interests as *women*, as *wives and mothers*, as *members of neighborhoods and communities*, and as *members of particular race, ethnic, and class groups*. To collapse a complex political consciousness into the more narrow confines of domestic values, interests, and roles is to distort both the motivation and the political implications of this mode of resistance (Morgen 1988, 111–112. Emphasis in original).

Through their participation in local grassroots associations, women create a legitimate voice to represent their interests in the governing bureaucracy; what one scholar calls "activist mothering" (Naples 1998 and 1992. See also Orleck 1997; Basu 1995; Taylor 1993; Phillips 1991; Sacks 1988; Parenti 1970). When such organizations are managed and led by women they often cultivate local women's leadership, provide greater opportunities for women's involvement, and enable participants to develop skills transferable to other forums. As an avenue for political participation, grassroots organizing has proven to be especially accessible to women, often leading to greater opportunities for leadership (McGlen and O'Connor 1995; Thomas 1994; Hardy-Fanta 1993; Cantor and Bernay 1992; Naples 1992; Morgen 1988; Boneparth 1984; Dabrowski 1983; Rendel 1981). Furthermore, and perhaps more important, women's political involvement creates a situation whereby

> ...women begin to see the contradictions in ignoring their own oppression while fighting against other injustices. ...Attacks by men (and women) who question women's intentions or ridicule their ideas and behavior can contribute to greater consciousness and solidarity among women activists (West and Blumberg 1990, 21).

It is estimated that 90 percent of those who actively participate in the environmental justice movement are women (Di Chiro 1992, 109). They are not, however, proportionally represented in highly visible leadership positions within the movement. Nevertheless, a great many environmental justice groups in the U.S. and abroad are founded and run by women of color (Di Chiro 1992, 109. See also Paehlke and Rosenau 1993; Boneparth 1984). As Di Chiro points out,

> what is 'new' about the environmental justice movement is not the 'elevated environmental consciousness' of its members but the ways that it is transforming the possibilities for fundamental social and environ-

> mental change through processes of redefinition, reinvention, and construction of innovative political discourses and practices (1992, 98).

These activists, who tend to work through structurally and ideologically democratic organizations to improve conditions in their communities (Mack-Canty 1997; Taylor 1993 and 1990; Cable 1992; Heiman 1990; Hamilton 1985), are often mothers fighting problems that threaten the health and well-being of their families and neighborhoods (Kaplan 1997; Rocheleau et al. 1996; Seidman 1994; Bullard 1993; Hamilton 1993; Moreno 1991; Taylor 1990; Hall 1988). Their motivation, however, is not solely a product of their biological or gendered socio-cultural roles of "mother" and "nurturer" (Di Chiro 1992).[5] Rather, their motives for involvement in such grassroots politics, and/or in electoral politics, are far more complex. As Di Chiro notes,

> the question of community survival in the face of cultural imperialist attacks by the dominant, white male, industrial complex figures conspicuously in many women of color's involvements in environmental justice works. Women in many communities and cultures have customarily been seen to be the repositories of or given the responsibilities for maintaining local, cultural traditions and histories (1992, 115).

By obtaining mass support and adopting active leadership roles, indigenous and Hispanic women in New Mexico's environmental justice movement similarly influence public policy and shape the politics of the state. As I will argue, the political activism of the indigenous and Hispanic women leaders I interviewed spans grassroots and electoral politics. While the New Mexico leaders' motives may incorporate traditional maternal and domestic concerns, the reasons for their political involvement is not limited to their gendered identities as women, wives, or mothers. Instead, their trajectories into politics originate from many sources, including their early political socialization, their particular experience of politicization, their affinity with and commitment to their racial/ethnic community, their identification with and desire to improve the condition of women as a group, and their political ideology.

We now turn to a discussion of the 50 grassroots activists and public officials interviewed for this project. Chapter 4 provides an introduction to this group of leaders examining their demographic characteristics, their experiences of politicization, their motives for participation, and their varied paths to political leadership.

Chapter 4
Women's Paths to Leadership

> Leadership is a complex phenomenon revolving around *influence*—the ability to move others in desired directions. Successful leaders are those who can take full advantage of their opportunities and their skills. Institutional structures, the immediate situation ... the political culture, regime type, and the dynamics of followership define the opportunities for the exercise of leadership. The leader's style, political acumen, character traits, and personal attributes provide a behavioral repertoire, a set of skills. Opportunities and skills interact to determine the success or failure of attempts to lead and influence (Genovese and Thompson 1993, 1-2. Italics in original).

While American Indian women and Hispanas remain underrepresented in New Mexico's political leadership, a few do hold high-profile positions in local, state, national, or tribal government (see for example Glover 1998; Jacquez-Ortiz 1995; Paul 1995; Contreras 1993b; Contreras and Shaw 1993) and in environmental justice organizations within the state (Diane 1995; Seidman 1994; Contreras 1993a; Conrad 1991; SouthWest Organizing Project 1991). As my research will show, despite their small numbers, these women participate in the public policy process to successfully effect change for their communities. I hope to begin filling some of the gaps in the literature by shedding light on the politics of these Hispanic and indigenous women leaders. By addressing the following research questions, this chapter provides an introduction to the New Mexico study participants and initial analyses of the data:

- Who are these political leaders? What are their characteristics?
- What were their experiences of political socialization?
- What were their motives for assuming leadership?
- How did they acquire their positions of political leadership?

The findings illustrate the diversity of these leaders and their varied political trajectories while revealing the prevalence of gender, race and ethnic identity as motivations for Hispanic and indigenous women's political participation in New Mexico's environmental politics.

WHO ARE NEW MEXICO'S INDIGENOUS AND HISPANIC POLICY LEADERS?

> I love politics. I love working with people and for people. It's a different way of giving back to your community. I feel a sense of gratitude and pride...knowing that you really do help people. ...It's a very fulfilling job. But I don't feel that the public really knows what you go through to get where you're at. And especially sometimes as women, we sacrifice so much. But we don't talk about it. We sacrifice how we feel sometimes. We sacrifice time with our families. We sacrifice our own personal time (Pilar, public official).

The 50 New Mexico leaders who participated in this study were extremely diverse and, in some respects, atypical of women in the general population with regard to their demographic characteristics.

Table 5. Age of Leaders in Study and Women in New Mexico

Age	Activists		Public Officials		Total Leaders		NM Females [1]	
	No.	%	No.	%	No.	%	No.	%
18 to 24	1	4	0	0	1	2	75,122	14
25 to 34	6	23	0	0	6	12	129,020	23
35 to 44	8	31	11	46	19	38	114,963	21
45 to 54	7	27	10	42	17	34	75,610	14
55 to 69	4	15	3	13	7	14	93,418	17
Total	26	100	24	100	50	100	488,222	89

[1] Females in New Mexico 18 years and older (N = 550,907). Source: U.S. Department of Commerce. *1990 Census of Population—General Population Characteristics, New Mexico.* 1990 CP-1-33 (from Table 28).

The activists and public officials ranged in age from 24 to 64; as Table 5 indicates, most (36 or 72 percent) were between 35 and 54. The median age for the Hispanic leaders was 45 years and for the Indian leaders, 44 years. Public officials tended to be older than activists, possibly because acquiring public office—either by appointment or election—requires considerable experience and time. Some of the public officials who started out as activists only entered electoral politics later in their careers. Others waited until their children were grown and/or out of school before they pursued public office. With a median age of 45 years for the activists and 50 years for the officials, the sample of leaders was somewhat older than the median age of 32 years for women in New Mexico.

In general, the level of education attained by the leaders also varied considerably from that of women in the state, as is shown in Table 6.

Table 6. Educational Attainment of Leaders and Women in New Mexico

Educational Attainment	Activists		Officials		Total Leaders		NM Females [1]	
	No.	%	No.	%	No.	%	No.	%
Grade 9-12, No Diploma	3	10	0	0	3	6	67,797	14
High School Grad / GED	3	10	2	8	5	10	145,350	30
Some Coll./Tech. Train.	4	17	7	29	11	22	125,858	26
Bachelor's Degree [2]	5	23	6	25	11	22	53,648	11
Master's Degree	4	13	1	4	5	10	24,547	5
Prof'l School Degree	2	7	6	25	8	16	4,566	1
Doctorate Degree [3]	5	20	2	8	7	14	2,279	0.05
Total	26	100	24	100 [4]	50	100	424,045	87 [4]

[1] Females 25 years and over in New Mexico (N = 479,406). Source: U.S. Department of Commerce. *1990 Census of Population—General Population Characteristics, New Mexico.* 1990 CP-1-33 (from Table 28).

[2] These figures include leaders with some post-graduate work but no Master's degree.

[3] This category includes leaders with two or more Master's degrees and leaders who were enrolled in doctoral programs.

[4] This amount is rounded.

Whereas only 16 percent of the leaders (6 activists and 2 officials) reported having between nine and 12 years of schooling, 44 percent of women in New Mexico had a high school diploma or less. Hispanic and Indian leaders both had a median of 16 years of education or the equivalent of a Bachelors degree. At the time of their interviews, several women in the leaders' group were fulfilling requirements for their Bachelors degree or taking graduate courses. Twenty or 40 percent had completed graduate degrees, a remarkable figure when one considers that only about 6 percent of New Mexican women accomplish this feat.

Of the 20 leaders with post-graduate degrees, 15 had completed a Juris Doctor degree, or a PhD, or more than one Master's degree, and one woman was enrolled in a doctoral program. Eight law degrees were evenly divided between the Indian and Hispanic leaders. And, since conducting these interviews, I learned that two of the activists (one Indian and one Hispanic) had entered law school. As a group, the Indian activists had the greatest number of post-graduate degrees (7), followed by the Hispanic officials with five. Four leaders in each of the remaining groups, the Indian officials and the Hispanic activists, had post-graduate degrees. On the whole, the New Mexico leaders were exceptionally well educated.

The leaders' generally high level of education, however, was not necessarily reflected in their household income, as we can see from Table 7, below. Despite their education, there was a broad range of leaders' incomes with public officials' households having greater incomes than those of activists. The data also revealed differences among the household income of the Hispanic and Indian leaders. The median household income for Hispanic activists was somewhat lower ($35,000) than that of Indian activists ($40,000), but Hispanic public officials[1] reported a slightly higher median household income ($61,000) than that of Indian officials[2] ($58,000).

Table 7. Household Income of Leaders

	Activists		Officials			
Household Income	Indig.	Hisp.	Indig.	Hisp.	Total	%
Up to $9,999	1	1	0	0	2	4
$10,000 to $34,999	2	4	3	1	10	22
$35,000 to $74,999	7	7	3	8	25	54
$75,000 to $99,999	2	0	2	2	6	13
$100,000 or more	0	0	3	0	3	7
Total [1]	12	12	11	11	46	100

[1] Of the 50 women interviewed, three chose not to reveal their income and one did not know her household income.

The median for all of the leaders, as a group, was $57,500. However, there was a tremendous range in the income reported by the leaders; three women had household incomes of $100,000 or more and two women had household incomes of less than $10,000. As one activist so aptly put it, "What I make is peanuts."

Table 8, below, provides a comparison of household incomes between the leaders interviewed and the New Mexico public. At first glance, the leaders appear to enjoy a higher income than most in the state. In contrast to 31 percent of the general public, 74 percent or 34 of the 46 leaders who disclosed this information, reported having a household income of $35,000 or greater. Once we examine separately the household incomes of those public officials and activists who volunteered this data (see Table 8), it becomes clearer that the household incomes of the activists are more similar to those of households in the New Mexico public, and less like those of the public officials. For example, while the majority (68 percent) of households in the state and 33 percent of the activists had incomes below $35,000, only 18 percent of the public officials reported household incomes in this range.

Table 8. Household Income of Leaders and New Mexico Public

Household Income[1]	Activists		Officials		Total Leaders		NM Households[2]	
	No.	%	No.	%	No.	%	No.	%
Up to $9,999	2	8	0	0	2	4	107,409	20
$10,000 - $34,999	6	25	4	18	10	22	261,984	48
$35,000 - $74,999	14	58	11	50	25	54	143,925	26
$75,000 - $99,999	2	8	4	18	6	13	17,069	3
$100,000 or more	0	0	3	14	3	7	13,438	2
Total[3]	24	100	22	100	46[4]	100	543,825	100
Median	$37,500		$52,500		$57,500		$24,087	
Mean	$40,000		$60,000		$49,000		$31,006	

[1] The amounts shown here are not entirely comparable with U.S. Census figures as my data are based on the leaders' household income for 1991, 1992, or 1993, depending on when their interview was conducted. However, the data do allow for some rough comparisons of household income between the leaders and public.

[2] All New Mexico households that reported income for 1989. Source: U.S. Department of Commerce. *1990 Census of Population—Social and Economic Characteristics, New Mexico.* 1990 CP-2-33 (from Table 28).

[3] Percentage totals are rounded to 100.

[4] Three leaders declined to disclose their income and one did not know her combined household income.

The difference among the three groups is highlighted when we consider that only 5 percent of the New Mexico public and 8 percent of the activists enjoyed a household income of $75,000 or more, in contrast to one-third (32 percent) of the public officials. The variance in the leaders' incomes may be attributed to numerous factors, including the difference between single- and multiple-income households, variations in educational attainment, and the fact that some women held well-paid full-time professional positions (primarily in the public sector), while others held poorly paid positions (mainly in non-profit organizations), were employed only part-time, or were retired from the paid workforce.

Several leaders held positions that paid poorly, whether as activists or public officials, but many of them had additional sources of income.[3] Both of the women who reported total household incomes of less than $10,000 were self-supporting activists whose primary work was their community organizing activities. While both had 12 years or less of formal schooling, not all of the leaders with modest household incomes had low levels of education. Two activists with Master's degrees reported household incomes of $16,000 or less. As one leader stated,

> It's hard to find people with the education that I have that are willing to take a low paying job like this...I know I could find another job that pays much better. But I don't want a job that has no real impact on the future of this country or this planet. We deal with life and death issues here. Everything we do has an impact (Rose, activist).

Many of the leaders interviewed, whether activists or public officials, had chosen to forego earning higher incomes (even though many had the necessary education to do so), in favor of committing their time to their political work. One leader accepted a lower paying position with a non-profit environmental group although her technical degree almost guaranteed her a high paying job right out of college:

> I...got other job offers around the country. My family expected me to make big money because I was the first one to graduate with a degree. I was sent a fact sheet from this organization but I thought "I'm overqualified" and threw it away. Then I realized this was the place to be. I could've been at DOE [U.S. Department of Energy]. The turning point [came when] board members here encouraged me. They recognized I have a commitment to the people of New Mexico, and that's what they wanted (Felicia, activist).

Making money was not the principal source of motivation for these particular women or their cohorts.

Balancing Family Life with Political Leadership

> I want to spend more time with my family. It's hard to balance out; I want to represent the people as much as possible and be with my family. I'm away [a lot]. Politics takes an incredible amount of time. It basically minimizes the time you can spend with your family. As much as possible, I try to involve my daughter in what I'm doing so that she doesn't resent my being away. Sometimes I can take my family with me but it's hard with a two-and-a-half year old (Gina, public official).

> I was married at 20. I wanted to be the perfect wife; to accommodate my husband—for 20 years. Then one day I woke up and realized I could manage myself. It gave me guts. [My community work] has changed my life. It's given me more self-assurance (Dalia, activist).

Women have traditionally been relegated to the domain of the home and hearth reducing women's realm of socially sanctioned activity to what has been commonly termed the domestic or "private" sphere by feminist scholars. By limiting their role to the private sphere, women are effectively excluded from politics, and their political participation is marginalized. It is therefore essential that, as Cohen, Jones, and Tronto assert, we

> provide a richer understanding of the complex contexts of [women's] political engagement—family and intimate relationships, friendships, neighborhood, community, work environment...race, religious, and other cultural groupings—that structure perceptions of opportunities for political participation (1997, 7).

Indeed, several researchers have documented how women in politics face multiple barriers unique to their sex such as the gendered stereotypes of "inherently" feminine qualities that supposedly make women less fit for the rigors of politics (Fox 1997; Deutchman 1992) and affect their success at, for example, fundraising for their own campaigns (Burrell 1994; Witt et al. 1994). Sex-role stereotyping has also resulted in fewer opportunities for women to practice formal leadership or to gain valuable political experience, to receive mentoring from politically powerful women, or to have female role models who are political figures (Carroll 1994; Genovese 1993; Cantor and Bernay 1992). This is especially true for women of color who must contend with racism in addition to sexism and may help to explain why only one-third of the New Mexico leaders interviewed were mentored, as well as why so many of the public officials were among the first Hispanic or Indian women in their particular leadership position.

Society continues to hold expectations about women's primary role and duty as wife and mother, which constrain women's ability to fulfill their own political ambitions (Kahn 1996; Lobao 1993). Consequently, women's political careers are often delayed because of family obligations, and, due to their late start in politics, women are left with less time to climb to the highest levels of leadership (Cannon and Marusich-Smith 1986). This may be evidenced in the difference among the median ages of women in the New Mexico public (32 years) in contrast with the median ages of the activists (45 years) and public officials (50 years).

Once in politics, women seeking to excel in this traditionally masculine arena often find themselves confronting obstacles, many of which relate to their role in the private sphere (see for example Hartmann 1987; Hartsock

1987). As Tronto notes, "while women vote in proportion to men, women are underrepresented in any other form of political participation, in large part because their private duties keep them out of public life" (1993, 175-76). Because domestic duties such as upkeep of the home, child rearing, and care of the family overwhelmingly remain the responsibility of women, those seeking alternative or nontraditional roles often face opposition. As one New Mexico leader confided, "My marriage fell apart right after my public and political involvement increased. I spent a lot of time away. You need a partner to support that" (Sonia, public official). Opposition to a woman's active political involvement may be subtle or overt, it may originate in the family or in the larger society. Regardless, women in positions of political leadership confront difficulties on account of their sex, which men, most often, do not. As West and Blumberg note of women who participate in grassroots politics,

> Women's engagement in social protest requires them to break out of informal customs as well as oppose the overt and formal power of patriarchal authorities in all institutions. ...Women are still sanctioned within gender-integrated movements and within the family for asserting their rights, regardless of their ideology. From this perspective, the costs of participation for women are clearly higher than for men (1990, 32).

In a 1987 study of women leaders in American politics, Le Veness and Sweeney identified the opportunity structures available to women who broke into top positions in electoral politics (also see S.J. Carroll 1994). The authors similarly found that these elite women generally grew up in families that both imparted strong values and supported their goals. Further, the women leaders had achieved levels of education on a par with men who had corresponding degrees of political success. And, like the New Mexico leaders, these women "learned that the behavioral rules imposed on female politicians are stricter than those imposed on male politicians" (Le Veness and Sweeney 1987, 4). Along the same lines, women in political leadership "are expected to defer to their roles as mothers and homemakers" (Le Veness and Sweeney 1987, 5). Consequently, many of the women had waited to enter public life until their children were older, as was the case for some of the New Mexico leaders.

Unlike the New Mexico leaders, however, the women in Le Veness and Sweeney's study varied substantially in their support and work on behalf of women's issues. The difference between the levels of support exhibited by these two groups of leaders may simply be a function of time as the New Mexico leaders were interviewed at least 10 years later, during the 1990s. Greater numbers of women have since entered politics, and as their presence has grown, so has their influence in public policymaking[4] (see

O'Regan 1997; Kahn 1996; Burrell 1994; Thomas 1994; Havens and Healy 1991; S.J. Carroll 1989b; Saint Germain 1989). Finally, like the women interviewed by Le Veness and Sweeney, the New Mexico leaders' trajectories into politics varied considerably.

Whether or not they personally experienced sexual discrimination, the New Mexico leaders maintained that women generally must work harder, and jump greater and more numerous hurdles than men, if they wish to attain influence and respect in politics. Research supports these findings (see for example Fox 1997; Burrell 1994; S.J. Carroll 1994; Holtzman and Williams 1989; Saint-Germain 1989; Cannon and Marusich-Smith 1986). Similarly, women tribal leaders in a national study perceived that

> a woman leader must be a "superwomen" [sic]—mother, wife, and tribal leader—while the man leader does not have to play all these different roles. ...[W]omen tribal leaders have to fit into the male world of tribal politics and not vice versa. Women leaders must "play the game" if they are to be effective. This "good old boy" network puts women at a disadvantage (McCoy 1992, 65).

Consequently, having support, encouragement, and material help from their mate and family can contribute to their effectiveness as the organizer of a grassroots campaign or the successful outcome of a woman's race for public office.

> I don't think anyone understands how difficult it is to run a campaign and be a single mother. It's almost impossible. And I only ran county-wide. But you have to raise money, you have to attend functions, you have to work during the day, and run a campaign, and then at the same time be mom. Even to get to events that I should be at, sometimes I just can't because of my responsibility as a mother. It's too much for one person. It's a lot of different focuses. You're all spread out, and to do that well, it's just really hard. Anyone who's smart probably wouldn't do it. [Laughs] ...It's very draining and it takes over your life. I came back into my life [after the campaign] and all of a sudden, 18 months of it were gone! You put off a lot of things because you just can't give it the energy or attention it needs. Having an extended family that's supportive makes it less of a problem (Tammy, public official).

On the other hand, resistance, opposition, and discouragement from mate, family, and/or friends may deter women from even attempting to involve themselves in public life. For some leaders, therefore, maintaining an active leadership role meant either finding an accepting (and preferably supportive) partner, or remaining single.[5]

> I saw how my mother's involvement put a strain on my parents' relationship. For me, the first criteria I had for a mate was that he'd be an Indian...very educated, and so secure in himself and his career that I would not be a threat to him. ...About seven years ago, I was very busy, gone a lot for meetings and work—it got really old for my husband. I asked "What do you want me to do? Quit?" He said, "All the qualities that I like in you are making you involved and are taking you away." After about two years, it was too much for me, too. I was happy but stressed out as hell. [So, I reduced my involvement]. I give a lot of credit to my husband. He cooks, cleans, picks up the kids when I'm gone. I couldn't do this without his support (Tanya, public official).

While some women in this group of New Mexico leaders attained their political position despite the opposition and/or lack of support of their families, most found the assistance and encouragement provided by others to be invaluable to their success. We now turn to an accounting of the conditions that contributed to and enabled the political involvement of these leaders.

As we can see from Table 9, the Hispanic and Indian women were remarkably similar in terms of their family make-up. Twenty-seven women, 54 percent of both the Indian and Hispanic leaders, had school-age children or a dependent elderly adult living at home at the time of their interview(s). Of those leaders with dependents, 24 or 89 percent had only one or two.

Table 9. Leaders' Family Status

Family Status	Activists		Officials		Total	%
	Indig.	Hisp.	Indig.	Hisp.		
Has Supportive Family[1]	11	11	11	9	42	84
Is Single	5	4	3	5	17	34
Has a Live-in Mate	8	9	10	6	33	66
Has Supportive Mate	6	8	10	4	28	85[2]
No Dependents at Home	5	4	7	7	23	46
Has Dependents at Home	8	9	6	4	27	54
1 Dependent	5	4	3	2	14	52[3]
2 Dependents	2	3	3	2	10	37
3 Dependents	1	1	0	0	2	7
4 Dependents	0	1	0	0	1	4

[1] This refers to whether the leader grew up in a family supportive of political involvement.

[2] This figure is calculated based on the number of leaders with live-in mates (N=33).

[3] The following percentages are calculated based on the number of leaders with dependents living at home (N=27).

The great majority of leaders, 42 or 84 percent, stated that their families were supportive of their political involvement. Being "supportive" meant that their children, mate, parents, and/or other family members actively sustained the leader's political involvement in one or more of the following ways: by working on her campaigns or other projects, helping her with household responsibilities, providing child care, and/or accompanying her to political functions. The following experience related by one woman was common to several of the leaders interviewed:

> My family has always been there for me, to take care of my kids or help in any way. During the campaign, they were all licking envelopes and passing out literature. ...I have great family support; you need it more as a woman (Gina, public official).

In many cases, leaders included their families in their political work to a great extent. For example, several women described how they brought their preschool children with them to meetings, or how older children helped in campaigns and petition drives.

> Our children have always been involved. Our daughter took her first steps in a public meeting. My husband stopped the meeting and everybody applauded. My son was two when we were in Taos and the FBI was taking photos. Our kids went everywhere with us (Lucia, activist).

The majority of the leaders, 33 or 66 percent, were married or had live-in partners. There was only a slight difference between the Indian and Hispanic women on this issue. Whereas 69 percent of the native women had live-in mates, 63 percent of the Hispanas did. Furthermore, of those married or partnered women, a slightly higher percentage of Indian (89 percent) than Hispanic women (80 percent) reported having a supportive mate. It is not clear whether more of the Indian women were partnered *because* they had supportive mates but it is significant that large numbers of both groups of women felt their partners were supportive of their political involvement. In fact, the majority of these women stressed the value of their mate's help and moral support to their success as political leaders.

The responsibilities associated with parenthood, maintaining a relationship, managing a household, and holding a job (either paid or volunteer) demand considerable time and energy. Receiving assistance from one's family and/or mate can determine a woman's ability to maintain a high level of political activity. As one seasoned Pueblo leader said of her husband,

> He's been my biggest supporter. I wouldn't have been able to do it without [him]. ...When I've been away on trips and stuff he's been there for the children. It's really been a godsend to have him. ...As far as he's concerned, there's no such thing as "man's work" or "woman's work." We've always worked together. When I'm late coming home from a meeting, he doesn't think anything of starting the supper or putting a load of clothes in the machine, whatever. Thank god, [laughs] I'm very lucky (Tracie, public official).

Only five of the 33 women who were partnered did not have supportive mates. But whether their partners were supportive or not, several women commented on the stress that their public involvement placed on their relationships with their mate and families. As one woman noted,

> I was a community organizer before I ever met [my husband] and he knew he wasn't going to change that. [Laughs, then pauses] It's hard on him to run the house when I'm gone. But [shrugs her shoulders] he doesn't try and stop me (Marta, activist).

In general, lack of material and/or emotional support from one's mate (or from her family) was seen by many of the leaders as a handicap faced by women in politics far more frequently than by men. In fact, studies indicate that women face certain barriers to their political involvement peculiar to their sex (Fox 1997; Burrell 1994; Cruz Takash 1993; Cantor and Bernay 1992; Antolini 1984). For example, they are more likely than men to (have to) seek permission from their mate, familial support, and/or assistance with their household duties in order to pursue political activities (Witt et al. 1994; Cable 1992; Cannon and Marusich-Smith 1986). On the other hand, a wife's support of her husband's political and/or career aspirations is nearly always assumed. And, indeed, this is evidenced in a statement made by a New Mexico Democratic party activist and former candidate for lieutenant governor:

> Women tend to be very methodical. They look at their support systems and ask, "Will I get the support I need from my family, from everyone else?" Men just assume the support will be there, and it usually is. Women find they don't always have the support they need (Diane Denish, quoted by Paul 1995, 9).

The burden on single mothers who lacked the support of their family was greatest as they struggled to successfully fulfill the demands of their many roles. For example, when asked "What gets in the way of your political work?" one woman replied, "My priority of raising my children. Taking care of my granddaughter for my daughter. Family concerns. I spread myself too thin" (Flora, public official). Flora was the head of her household and primary earner. She had guardianship of her granddaughter, held a full-time job, was a leader in her labor union, and served on her local school board. Her multiple commitments and responsibilities were not unusual for the group of leaders interviewed, however. Where family members were non-supportive, some leaders were forced to solicit help from friends.

Conflicts between the domestic role of women and their desire to be active in politics may be less of a problem, however, for this group of New Mexico leaders. While many of the women interviewed were mothers and/or wives who felt the strains of family obligations, the vast majority reported having supportive families and mates who frequently assisted with and participated in the leaders' political activities. Those who felt the greatest conflict between their traditional role as mother and their role as grassroots activist or public official were, understandably, single mothers whose families were not supportive of their political activism. These women sustained themselves and their children through their strong will, hard work, and their belief in themselves. Their confidence and determination was evident in their self-image.

Leaders' Self-Perceptions

The leaders interviewed characterized themselves in a rich variety of ways, such as "open-minded," "a risk taker," "an activist," "an advocate," "independent," "spiritual," "caring," "a role model," and "hard working." Nearly half of the women (22 or 44 percent) referred to their sex or gendered roles when describing themselves as a "woman," "mother," "parent," "wife," or as "family-oriented." The prevalent themes in their narratives included service, responsibility, challenge, commitment, and culture. References to their cultural and racial/ethnic identity were mentioned by 20 or 40 percent of the women.[6]

The majority of the leaders (64 percent), in contrast to 36 percent of the New Mexico public, were either bilingual or multilingual in English and their native language(s) (see Table 10 on the following page). Of those who spoke a language other than English at home or while growing up, 47 percent of the leaders and 79 percent of the general public claimed Spanish as their native tongue. Native North American languages were spoken by 53 percent of the leaders and 16 percent of the public who spoke a language other than English at home or while growing up. While several leaders said that they understood Keres, Tewa, or Spanish, for example, unless they were conversant in that language, they were not considered bilingual for purposes of this study. Many of the indigenous women who did not speak their traditional language(s) expressly stated that as children, they had been discouraged by their families from speaking a language other than English.

Table 10. Native Language of Leaders and New Mexico Public

Native Language[1]	Activists		Officials		Total Leaders		NM Public[2]	
	No.	%	No.	%	No.	%	No.	%
No. Amer. Langs.[3]	8	31	9	38	17	34	79,087	6
Spanish	9	35	6	25	15	30	388,186	28
Other Languages	0	0	0	0	0	0	26,726	2
Total Bi-/Multilingual	17	66	15	63	32	64	493,999	36
Total English Only	9	35	9	38	18	36	896,049	64
Grand Total	26	100%	24	100%	50	100%	1,390,048	100%

[1] Whereas I asked the leaders which language they grew up speaking at home, the U.S. Census Bureau asks for the "language spoken at home."

[2] All persons in New Mexico five years old and over (N = 1,390,048). Source: U.S. Department of Commerce. *1990 Census of Population—Social and Economic Characteristics, New Mexico.* 1990 CP-2-33 (from Table 18).

[3] Native North American languages spoken by the leaders include: Comanche, Dine (Navajo), Keres, Kiowa, Shoshone, Tewa, Tiwa, Washoe, and Zuni.

> I understand a little bit of Mohawk. I wasn't brought up traditionally although my parents both speak our language fluently and were brought up that way. They didn't bring up their children that way because they thought it was a disadvantage. They moved off the reservation so I grew up in white culture, although I always knew I wasn't white. I don't think it impeded me or got in my way. But they wanted us to do well, to be self-sufficient, and we all are. I guess their plan worked to that extent. Although I wish that I knew my language and I don't (Kari, activist).

In most cases, it was thought that speaking their parents' language would interfere with their ability to speak English, potentially harming their opportunities for success in the dominant Anglo culture.

> My parents were both professionals, they spoke English at work and when they came home. They consciously didn't speak Keres because they didn't want us [children] ostracized. I picked up Keres from listening to my grandparents and relatives [in our Pueblo]. I'm not as fluent as I can be. I can get by to do what I need to for religious things (Tanya, public official).

In general, a leader's sex, her gendered identities, her connection to family and community, culture, and race/ethnicity all served to inform and shape her participation and political leadership. The importance of gender and racial/ethnic identity will be explored in greater detail in the following chapter. The remainder of the present discussion will focus on (1) the role of the family on the participation and political socialization of these leaders, (2) their motives for entering politics, and (3) their trajectories into positions of leadership.

LEADERS' EXPERIENCES OF POLITICAL SOCIALIZATION

> I've always been in leadership roles. My grandmother used to say: "Here comes the boss" (Victoria, public official).

A pattern of life experiences pertaining to the leaders' early socialization, community involvement, and current domestic situation emerged from the data. These experiences, which many of the leaders had in common, contributed to their politicization and initial participation in public life.[7] Other factors, such as their present family life (or their ability to succeed in spite of it), enabled them to maintain a high level of political activity.

The Role of Politically Active Families

> When I was 8 or 9... there was a Navajo Councilwoman who was my role model. ...My parents...always sought her advice. She was very strong, very open. Helping her people. She has stuck in my mind. I try to model myself after her. She was assertive, not afraid to speak out (Flora, public official).

The majority of the leaders interviewed were socialized early on to actively engage in their community's affairs. Many grew up in households where either their parents or other family members participated regularly in public service, and the importance of civic responsibility was stressed.[8]

> My family shares, talks about politics. It's part of our value system. It makes the family stronger. ...Our mother was as involved as our father. With us, it's like running the family business, everybody was engaged in a real way (Cora, public official).

A closer examination of their introduction to politics and the breadth of their civic involvement helps us to understand these womens' motives for participating in public life.

Table 11, below, indicates that 64 percent of the leaders reported being taught civic values in their home. Specifically, 58 percent of the Hispanic leaders (14) and 69 percent of the Indian leaders (18) were raised in families who taught them the value of responsibility, service, integrity, and commitment. As one leader recalled,

> I remember being a little girl and going around with my great aunt in Torrance county and Northern New Mexico to campaign on behalf of people... She'd go from ranch to ranch to talk about it. I enjoyed that. But my personal involvement, when I knew I could make a change, was in eighth grade [when I ran for student council president.] ...All the time when we were growing up, no matter what our successes, we were always made known that we had to give back to our community. My parents were always active in our school functions or in our church. I have to give them credit, they were our examples (Sonia, public official).

The majority of the leaders (68 percent) had politically active families. In particular, 63 percent of the Hispanas and 73 percent of the Indian women grew up in families that were highly involved in the politics of their community.[9]

Table 11. Factors Influencing Leaders' Political Socialization

	Activists		Officials			
Influential Factors in Youth[1]	Indig.	Hisp.	Indig.	Hisp.	Total	%
Was Taught Civic Values	9	6	9	8	32	64
Had a Politically Active Family	7	5	12	10	34	68
Had Mentor / Role Model	3	2	5	6	16	32
Active in Service Orgs.[2]	3	6	6	2	17	34
Active in Electoral Politics[3]	4	2	4	8	18	36
Active in Social Movement(s)[4]	4	9	3	3	19	38

[1] By "youth" I mean that time between childhood and one's early twenties.
[2] A leader may be active in more than one political area simultaneously. Service organizations included, but were not limited to: high school and/or college student government, civic groups, and clubs dedicated to providing community service.
[3] Electoral politics refers to participation in party politics—whether in one of the two dominant parties or in youth groups affiliated with political parties such as the Young Democrats, and/or local non-partisan or tribal politics.
[4] Social Movements refers to participation in such organizations as the Civil Rights movement, the Chicano movement, La Raza Unida Party, the American Indian Movement, the Peace movement, and labor unions such as the United Farm Workers.

Some leaders had relatives in elected or appointed office at various levels of government. Others had family members outside of the formal political arena who were, nevertheless, respected opinion leaders such as union organizers, social movement activists, community elders, Clan Mothers, or Medicine Men.

> My mother was very involved in community service when I was growing up. I remember going to meetings in [the pueblo]—the North American [Indian] Women's organization, the Council of American Indians, the school board. That's been my whole life. I went to college during the 1970s, there was Vietnam... Alcatraz and Wounded Knee. I did a couple of demonstrations then. I was very involved in the [Native American student organization] at [the university]. A main member [of the American Indian Movement] knew my mother real well (Tanya, public official).

In a few cases, both parents were political leaders in their own right. One woman's mother was precinct head of one political party at the same time that her father was precinct captain of the other party! Another leader's father was a member of the U.S. Congress while her mother headed a national civil rights organization.

While there were differences among the leaders' household income and level of education, as noted earlier, these demographic characteristics were less usefull indicators of social class than determining whether the women were raised in politically influential families. To a large degree, power and prestige or social class are determined by one's distance from the center of power and by their influence in decision-making processes. Therefore, to the extent that income levels vary tremendously depending on one's geographic location (especially between urban and rural communities), income alone is a poor indicator of social class. To put it another way, the range in income among residents of an Indian reservation or rural Hispanic village may be slight although the degree of influence they exercise in the community may vary considerably. Likewise, an individual from a working-class background may become upwardly mobile as their income rises, while a highly educated middle- or upper-class individual may earn a modest income. So, in and of themselves, income and education are less revealing of one's membership in the social and political elite than is their status as "insider" or "outsider." Those leaders who came from politically active families that were influential in their community clearly had the advantage of "insider" status over those women who were not socialized by their families into politics but had to make their own way. In this case, having had a mentor or insider contacts might have eased their entry into political life.

In fact, although most of the leaders grew up in politically active families, only 32 percent reported having been mentored. This may have to do with the mature age of many of the women interviewed and the fact that 20 years ago there were very few women in political leadership after whom they could model themselves. Nevertheless, 33 percent of the Hispanic women and 31 percent of the Indian women reported that they were encouraged by, and/or modeled themselves after, a politically active individual whom they admired. Role models and mentors included mothers, fathers, grandparents, siblings or other family members, teachers, friends, and community leaders. Several women stressed the important influence their mentor had in motivating them, in increasing their confidence, and in helping them to achieve their potential. And, it is worth noting that even now, mentors continue to be valuable allies and sources of inspiration for some of these leaders. This helps to explain why so many leaders recognized the impact that they themselves had as role models and mentors in their communities.

> One of the...things I try to do is to [mentor], advise, and offer encouragement to other women to take the lead, to speak... I'll say "it's your turn." I try to do this in my day-to-day work, otherwise people rely too much on me to take the lead. Women are held back by how the world defines them. Lots of women out there are developing leadership. My mom and my grandma [were my] role models. Though traditional, they are also non-traditional. My grandma worked outside the home, she had to to survive. She'd say "You don't need to depend on men. Go to college, don't get married" (Esperanza, activist).

Race/ethnicity and gender figured prominently in leaders' decisions to mentor. As one elected official explained, "Native American kids really lack women role models. It motivated me to be a role model" (Flora, public official). Another woman acknowledged,

> I'm probably some kind of a role model. I think that's important. Young Hispanic girls should feel that they can be professionals; that they are not cut off from opportunities and alternatives other than the traditional ones that young girls are sometimes made to believe they should follow (Tammy, public official).

Regardless of whether they themselves had been mentored, the leaders recognized their potential influence as role models and considered this to be both an honor and a responsibility. As another woman related with great relish,

> I was told, "A woman will never win this office." ...I defeated the male incumbent in the primary! This had a terrific impact. This year I gave the keynote address at New Mexico Girls' State. I told them about my campaign. ...They were interested in how to balance a family [with political office]. They were real motivated to pursue public office. I'm a role model—I hope to do a lot of good. They can see that a woman can hold this [type of] position (Gina, public official).

Several women created formal mentoring opportunities within their sphere of influence to prepare young women for future leadership: "I have student interns. I sponsor women to the New Mexico military institutions. Mentoring younger women is very important" (Sonia, public official).

The Role of Early Civic Involvement in Political Participation

Women's political participation often begins with community volunteer work and centers around local issues. And in fact, all of the leaders interviewed in this study engaged in public service. Whether grassroots activists or elected- or appointed officials, the majority performed services for their neighborhood, local school, tribe, or state. Those women working as staff in their grassroots organization and those who held paid full-time positions in the public sector also engaged in community activities outside of their employment. Many volunteered in some capacity. Examples of their volunteer functions included: serving on the board of their neighborhood association or on some other civic organization's board; donating services to a non-profit or public sector agency; or serving on a governmental commission or board.

Volunteer work fulfills an important obligation to the community and provides leaders with new knowledge and skills. Much of this activity may be directly related to the leaders' political work and to pet projects or programs, which they seek to promote. The interaction with their constituents increases the leaders' contacts, visibility, and influence while enhancing opportunities for coalition building.[10] Finally, volunteer work provides leaders with a sense of satisfaction and connectedness to their community important for their own morale. As West and Blumberg note:

> For women throughout the world, service and politics have been deeply interwoven and often related pragmatically in the home, in the church, in the schools, and in places of work. ...Any analysis of women's political lives must, therefore, consider both the service and political dimension of voluntarism (1990, 29. See also Dabrowski 1987).

Of the 50 New Mexico leaders, a full 70 percent (35) were involved in service groups, social movement organizations, and/or electoral politics as young women (early twenties or younger). Approximately one-third of the Hispanas (33 percent or 8) and Indian women (35 percent or 9) reported being active in service organizations during their youth, as were nearly equal numbers of the grassroots activists (9) and public officials (8). These service-oriented groups included but were not limited to: high school and/or college student government, women's and students' organizations, professional associations, civic and religious groups, and clubs dedicated to providing community service.[11]

> When I was in high school, there was a Catholic social action organization called the Young Christian Students and I joined that. One of its purposes was to get people to express their spirituality through engaging in social issues... That led to my staying with the YCS when I was in college, which in the early sixties was grappling with the issues...of racism, sit-ins, things like that. I was invited to join the national staff of that organization right out of college. Two years after I left college, I got invited to work for the Student Nonviolent Coordinating Committee. I stayed with them for seven years (Marta, activist).

Some of these service groups such as the Student Nonviolent Coordinating Committee also fell into the category of social movement organizations, which in the case of SNCC played an important role in developing leadership and broader based support for the Civil Rights movement.

Over one-third of the New Mexico leaders (38 percent or 19) were involved in social movement organizations during their youth including the Civil Rights movement, the Chicano movement, La Raza Unida Party, the American Indian Movement (AIM), the Peace movement, the Environmental movement, and labor unions, especially the United Farm Workers.[12]

> I've always been involved. In college, I was on the support committee for the United Farm Workers' boycott. We organized the picketing of Safeway stores in Seattle. César Chávez was one of my heroes (Esperanza, activist).

Whereas half of the Hispana leaders (12) were involved in such groups, 27 percent of the indigenous leaders (7) were mobilized by these organizations.[13] The objectives, scope, strategies, and tactics of social movement organizations were distinctly different from both those of community service groups and groups linked directly to electoral politics. The social movement organizations were principally involved at the grassroots with effecting social change through various means including holding public forums and workshops for community members, distributing information, and using direct action tactics such as demonstrating, picketing, and boycotting. Not surprisingly, twice the number of grassroots activists (50 percent or 13) than public officials (6 or 25 percent) reported participating in social movement groups during their high school or college years. It is important to remember, however, that like Jacinta, numerous public officials started out as grassroots activists.

> I've been involved in politics since I was in undergraduate school. Probably since the late sixties, early seventies. ...There was a lot of things happening on campus at that time. A lot of political movements.

> The Chicano movement [which I was involved in] was very heavy in the Southwest and the West and, you know, [I] did a lot of work with...Blacks and Latinos and Native Americans from all over the country. The politics wasn't [Party politics]. It was more a social political movement that I was involved in. I would go to some of the Democratic Party functions but I wasn't really involved in the Democratic Party. It was only later on that I [was involved in the party]...in the early eighties... (Jacinta, public official).

Finally, 36 percent of the New Mexico leaders (18) reported first becoming involved in electoral politics through political organizations during their youth. Of these women, eight were indigenous (31 percent) and 10 were Hispanas (42 percent). Their involvement variously incorporated participation in party politics—whether in one of the two dominant parties, in youth groups affiliated with political parties such as the Young Democrats, in local non-partisan politics such as municipal elections, or in tribal politics. Not surprisingly, twice as many public officials (50 percent or 12) as activists (23 percent or 6) reported being involved in electoral politics during their youth. Through their involvement in electoral politics, these women sought to affect governmental leadership by campaigning for candidates or referenda, fundraising, registering voters, and lobbying influential policymakers.

The younger women participating in this study were school age during the turbulent 1960s and 1970s, while the older ones were nearing middle age.

> Politically, I came of age in the 1960s during the Black, Third World, and Women's movements. I was 18 in 1966, at the height of the Vietnam war so I adopted all of that sixties rhetoric. We could change the world. ...I was 20 when the [native] renaissance movement began. It was an outgrowth of the Third World movement. I was recruited...as part of an attempt to begin an indigenous people's movement... We began the land rights struggles to recover land and to recover resources and advocate for native rights. Advocate language...cultural practices that were being forbidden, lost or destroyed. I've been involved in environmental issues since the 1970s because of...development... First it was agriculture, then the military, and tourism—they've destroyed our ecosystems. I've been [involved] ever since (Linda, activist).

There appears to be an age cohort effect. Whether they marched on Washington, D.C. or the state capitol, or organized residents locally, the period of the 1960s and 1970s was a significant time in the political development of many of the leaders who were already active.[14]

> When I was young in the sixties, you just had to be involved in social issues. Politics was one way... to influence the system and to make your concerns known. And to have some influence over the governance and the social policies of the country (Jacinta, public official).

What is significant about these findings is the tremendous scope and breadth of the participation of the New Mexico leaders in public life. Their involvement in service groups, social movement organizations, and electoral politics all provided these women with valuable organizational experience, influential contacts, and opportunities for leadership training.

> I've always been involved. My parents, my grandparents have always been involved. From my earliest recollections, it's always been there. I've worked on Navajo tribal campaigns. Also on state and national election campaigns. It's the orientation of our family: get involved to give back. You work on yourself in order to become a healthy, contributing person. I don't separate politics. For me, life is politics. It's negotiating. It's part of my commitment as an individual... (Charlotte, public official).

All 50 leaders reported having worked in some capacity as a volunteer in their community—either in their youth or as mature women. These women's experiences of public service—defined here as either paid or volunteer work performed as a service to the community—varied considerably as indicated in Table 12.

Table 12. Length of Leaders' Public Service

	Activists		Officials			
Years Active	Indig.	Hisp.	Indig.	Hisp.	Total	%
2 to 5	2	4	1	0	7	14
6 to 9	3	1	3	0	7	14
10 to 19	3	0	4	4	11	22
20 to 29	4	4	3	3	14	28
30 to 39	1	3	1	3	8	16
≥40[1]	0	1	1	1	3	6
Total	13	13	13	11	50	100

[1] Two Public Officials and one Activist responded "All my life" when asked how long they'd been involved in community service. The length of their public service was 48 years, 49 years, and 48 years, respectively.

While the median number of years spent providing public service was 16 for all of the leaders, as a group, their experience ranged from a minimum of two years to a high of 49 years (one woman's lifetime). Fifty percent of the leaders had been active in their communities from 10 to 29 years, a significant period of time. About one-third of the women (28 percent) had been active for a range of two to nine years, and 22 percent had been serving for 30 years or longer.

Hispanic leaders had generally been involved in community service longer than Indian leaders. Those with the lengthiest tenure were Hispanic public officials. This group had a median of 25 years of public service ranging from a low of 11 years to a high of 49. Hispanic activists followed close behind with a median of 23 years. Indian officials and activists had a median of 17 and 15 years of service, respectively. Considering the similarities in the demographics and living situation of the Hispanic and Indian leaders, the reason for the Indian women's somewhat shorter period of involvement is not entirely clear. We can infer, however, that unlike the majority of their Hispanic counterparts who were actively involved in civic and political organizations from their youth, the Indian leaders entered public life later on, as mature women. Overall, the New Mexico leaders exhibited a long-term dedication to service as most (36 or 72 percent) had been active on behalf of their communities for 10 years or longer. This dedication is exemplified by three leaders in particular who reported having been active in their communities their whole life. Their accounts of life-long civic involvement are meaningful in that they reflect a value system in which commitment to public service is an intrinsic human obligation, rather than an avocation.

MOTIVES FOR LEADERS' POLITICAL PARTICIPATION

> [Politics is] making a better world for your grandchildren...making it a better place for everybody. Not just being a taker in society but putting back something. I've really enjoyed it; it's been very rewarding. I will always be a politician or community activist, it's in my blood (Alicia, public official).

The activists and public officials participating in this study expressed many of the same motives for initially becoming involved in politics and for remaining active. For example, 28 percent or 14 of the activists and 42 percent or 21 of the public officials tied their participation to a sense of civic obligation: the desire to help others or to give back to the community.

Table 13. Leaders' Motives for Political Participation

	Activists		Officials			
Motive for Participation[1]	Indig.	Hisp.	Indig.	Hisp.	Total	%
Civic Obligation	9	5	12	9	35	70
To Empower Others	11	11	11	7	40	80
A "Calling" or Mission	3	4	3	4	14	28
Cultural Preservation	13	12	12	7	44	88
Personal Experience/Rage	5	5	4	2	16	32
Health/Environment	12	13	12	9	46	92
Advance Children's Issues	5	10	13	8	36	72
Advance Women's Issues	2	7	7	8	24	48

[1] Leaders could indicate multiple reasons for their political participation. The reasons are not ranked in any way.

This dedication to helping others through voluntarism and thereby improving social and economic conditions for the group has been similarly noted in studies of African American women's community-based activism (Kaplan 1997; Hall 1990; Hamilton 1989), in Hispanas' community service activities (Pardo 1998, 1995, 1990; Hardy-Fanta 1993; Hewitt 1990), in Native American women's social and political work (Hoikkala 1995; Ford 1990; Koester 1988), and in Euro American women's grassroots organizing (Naples 1992; B.A. Carroll 1989; S.J. Carroll 1989b; Ackelsberg 1988 and 1984; Dabrowski 1987 and 1983).

There is at least one significant difference between the political participation of women of color and that of Euro American women that deserves mention here. The racial and ethnic identity of women of color often plays a critical role for which there is no parallel in the case of white women. A white woman may enter politics as an individual for whatever political reasons. Because she is a woman, she may be perceived as advocating "women's issues," whther she does or not. A woman of color, on the other hand, carries a double burdern. She is assumed not only to repreent the interests of women, but to advance the political agenda of her particular racial/ethnic group – regardless of the fact that there is no single uniform agenda for either women or racial/ethnic minority groups. However, as long as women remain a minority in politics, they will be seen (to varying degrees) as tokens and as representatives of all women. This burden on women of color was particularly evident from the interviews with the New Mexico leaders, most of whom expressed an overwhelming sense of duty to their community.

While a majority of the leaders (70 percent) believed they had a personal responsibility for maintaining the well-being of their families, and by extension, their communities, a larger percentage of Indian women (81 percent) than Hispanas (58 percent) claimed to be motivated by their civic obligation. This finding is paralleled in McCoy's study of women tribal officials, where the majority of the leaders interviewed equated politics with public service and civic duty, and viewed their role as reformers who sought to "make things better" (1992, 62). As one of the New Mexico leaders simply stated, "Politics is about making a difference" in the world (Estela, public official). In fact, numerous leaders saw politics as a way of facilitating social change: "Politics is a way to change systems, institutions, and policy for the betterment of society" (Sonia, public official). And, as one native activist-turned-politician noted from her own experience, "Working the system can be more effective than attacking it" (Cora, public official).[15]

A large majority of the Indian women (85 percent) and Hispanas (75 percent), as activists (85 percent) or public officials (75 percent) were motivated to participate in community life by the opportunities they had for empowering others. One definition of empowerment is

> ...the ability to act with others to do together what one could not have done alone. Empowerment stresses the cooperative dimensions of human interactions and seeks to engage our imaginations, extend our potentialities, enable us as collective actors, ultimately to enrich our lives (Ferguson 1987, 221).

Many of the leaders felt responsible for assisting others—barrio residents, tribal members, other women, youth and/or elders—to find their own voice, to exercise their rights, and to become active players in the life of their community. Leadership, for these women, generally meant facilitating others to achieve a prominent role in determining their own destiny through participation in public decision-making. Both activists and public officials employed diverse strategies for capacity building and leadership development to accomplish this goal. For example, one activist was employed as a union organizer and volunteered her free time to an environmental justice organization. She found opportunities to empower others in both settings and included her sons in these activities in order to empower them.

> [I] organize the unorganized .. .empower the workers to represent themselves ... educate them to understand their rights. To ... defend themselves in contract or workplace violations. I guess my coworkers see me as a leader. When there is a problem, they look to me or they ask "What's she going to do?" I get along well with my coworkers. We share ideas back and forth. We're able to work together and support each other. I have both a formal and informal leadership role. I'm not

> afraid to talk back, raise concerns about issues. ... I participate in challenging the status quo, participate in challenging decisions that may not be just for the community or just for the workers. Participate in educating people around those issues, in moving people into action around those issues, empowering them to make a change (Esperanza, activist).

Ferguson advances the notion that women generally conceive of power as empowerment, as relational, and as lateral rather than hierarchical because "women's experience stresses continuity and connectedness with others and acknowledges reciprocity of needs" (1987, 221. See also Hardy-Fanta 1993; Sacks 1988; Deutchman 1986). She envisions women's leadership, therefore, as

> more of a pedagogical and facilitative activity; rather than standing at the top of the hierarchy, [feminist] leaders...stand at the center of the web ... stimulate and coordinate the activities of others and serve as a focus for marshalling the energies of the whole (1987, 221).

This particular analysis is consistent with my findings as well as with those of other researchers examining women's political activism (see Naples 1998a and 1998b; Kaplan 1997; Dodson et al. 1995; Duerst-Lahti and Kelly 1995; Klein and Ackerman 1995; Bystydzienski 1992a and 1992b; Albrecht and Brewer 1990; Hamilton 1989; Ackelsberg 1988). For example, in a study of appointed state officials, the majority of the women interviewed indicated that women approach power differently from men and that "women emphasized power-sharing and involving their employees in decision-making" (Havens and Healy 1991, 66). Similarly, the New Mexico leaders conceptualized power in terms of their ability to share their knowledge, experience, and influence with others for the purpose of empowering others. For the most part, this group of New Mexico women were inspired and/or driven to participate in politics by their belief in the interdependence and connectedness of human beings, by the value they placed on inclusivity and democratic participation, and by their desire to make the world a better place.

Rather than seeking personal gain, prestige, or financial rewards, these women expressed great concern for, and commitment to, the advancement of particular groups and/or communities. In fact, several women referred to themselves as leaders only reluctantly, preferring the terms "facilitator," "organizer," or "problem-solver" to describe their role. But however humble some of these women appeared to be, they were by no means naive about their influence or ability to exercise power. In general, the leaders recognized their capacity to effect change and secure resources for their constituencies or communities. As a Pueblo woman explained, "I speak for people who can't speak for themselves: children, Indians, the poor" (Tanya, public official). This sentiment is also reflected in McCoy's study of women tribal leaders who expressed the belief that

> Women listen more, are more objective, and try harder to get everyone's viewpoint than do men. Women leaders see themselves as there to serve the people. ... [W]omen leaders tend to deal more with people's problems. Women leaders sit and talk and compromise ... (1992, 65).

Tanya, and others, held that divine powers had bestowed them with gifts such as intelligence, strength, opportunity, and empathy for the purpose of entrusting them with the duty to represent politically and socially marginalized groups. Six (23 percent) Indian women and eight (33 percent) Hispanas saw their political involvement as a spiritually motivated mission or "calling" and spoke of the importance of their spirituality in their lives.

> I've chosen this. The first half of my life was dedicated to my family. And now I'm saying I want to do this for me. It's my calling. I'm on a mission from God. I love my work. I've never been so happy—physically, spiritually, emotionally. It's so fulfilling (Dalia, activist).

Kaplan (1997) similarly documents the centrality of religious faith to African American and black South African activists who struggle for environmental and social justice and whom she refers to as "prophet leaders." For 14 of the New Mexico leaders (28 percent), faith provided an invaluable source of support and strength, helping to guide them through tough decisions and maintaining them through difficult times.

Among the New Mexico leaders, those who were the first woman, Indian, or Hispanic to serve in their specific leadership position saw themselves as having an enormous (and somewhat burdensome) responsibility to other women and to their ethnic group. This obligation, however, was also valued as a tremendous honor. Valerie, the first woman to hold a particular position in governance in her tribe, firmly believed that it was her destiny, recounting how friends and family had predicted this from childhood. Further, she saw her formal leadership role, which was both religious and political, as part of a greater plan that extended beyond community and Pueblo, throughout time.

> We've always been survivors. Obviously, we're still here. In spite of all the attempts to get rid of us, we're still here. We're gonna be here. I think we have a task in the overall scheme of things, universally, that is gonna continue. The work's gonna continue. That's what keeps me going, is our children. ... It's a very spiritual thing for me. ...That's why

> I think my becoming _____ was just meant to be for whatever reasons. All those experiences I had when I was growing up as a child was preparing me for this 'cause I needed to have those experiences. I needed to have those skills before I could do what I did... [The] prophecy came true (Valerie, public official).

This Indian leader's narrative illuminates her profound sense of the constancy, strength, resistance, and resilience of her people through time; the predominance of spirituality in her life; and her determination to serve faithfully the interests of her pueblo. Similar sentiments were expressed in varying ways by many of the women interviewed.

Racial and ethnic identity figured prominently in leaders' motives for assuming influential decision-making roles as they sought to preserve water and soil quality, to revive centuries old artistic or agropastoral traditions, to repatriate native lands or artifacts, to prevent the commercialization of indigenous cultures, and to politically empower their peoples. Cultural preservation provided 25 of the 26 Indian leaders and 19 of the 24 Hispanas with motivation for their political involvement. As one official explained,

> [I serve] as a Native American voice in the [non-Indian legislative body] to educate [them] on Native American issues/matters. To educate them on who we are, how we think and feel so they can understand what normally isn't understood [by non-Indians]. ...I involve myself in issues affecting Native Americans outside my district. I have to give them a helping hand, support, or they'll lose out. ...I try to educate those not feeling the emotions. I also try to educate tribal people. ...I have a commitment to helping the tribe get better organized, locating resources. A commitment to my constituents who lack knowledge, the know-how. ...I want to ensure that policies don't discriminate. [I want] fair access to opportunities—not just money, to help us grow, be better citizens (Leonora, public official).

As illustrated by Leonora's comments above, her task was formidable; while she felt herself to be a lone native voice on the legislative body, she took responsibility for making the case for Indian peoples, and endeavored to elicit empathy and understanding from non-natives about Indian issues. Furthermore, Leonora sought justice while simultaneously working for empowerment and for substantive gains in programs, policies, and resources for tribes in and outside of her district.

Like their native counterparts, the Hispanic leaders were also motivated to participate in politics by the goal of preserving traditional New Mexico culture and practices for future generations, despite the tremendous development pressures facing economically depressed communities.

> In northern New Mexico we have one of the few pastoral cultures left in this country. And when a pastoral culture loses its land it becomes an artifact. And I was really very worried that with development pressures, people were going to have to sell their land. ...Can anything be done to bring rural residents together around not losing their culture and their resources as opposed to desperately saying "yes" to jails, to radioactive dumps, to resorts, to whatever? A lifelong question of mine is "How do you even the odds for rural residents that have these cultural relationships and that we're losing in this country?" We're losing that diversity, and the science, and the arts that go with it (Marta, activist).

Marta initiated a highly successful economic development project, which was owned and operated by community members based on traditional agropastoral practices. Besides invigorating the local economy, the project had a significant and positive impact on the lives of participants. Like Marta, leaders found that working to achieve cultural preservation often resulted in additional benefits such as the empowerment of residents, increased autonomy in decision-making, and material/economic benefits to the community. Nearly all of the activists in this study (96 percent), and a majority of the public officials (79 percent) were motivated to participate in politics by the goal of maintaining the integrity of their culture, language, and lifeways. We will return to the discussion of racial and ethnic identity in more depth in the next chapter.

Characteristic of the environmental justice movement, numerous leaders identified relationships between people of color, their land and labor, and environmental health issues. For 35 percent of the Indian leaders and 29 percent of the Hispanas, participation in politics stemmed from personal experiences that politicized them or sufficiently enraged them to mobilize for justice or social change. One leader was mobilized into action against the company that had mined her tribe's land for years, leaving the water and soil severely contaminated. Furthermore, she believed that the industry's mining of uranium on the reservation was directly responsible for the unusually high rate of cancer in the community. A cancer survivor herself, and previously apolitical, Donna had begun to make connections between patterns she saw in the siting of hazardous facilities and native lands.

> Everywhere you look there are all kinds of Native American Indians... dying of cancer... My question is why? Why are these companies trying to store this stuff on our reservations? ...Are they trying to kill us all? What are they trying to do? They have this slogan of "Keep America Beautiful," then why are they trying to destroy it? We Native Americans are taught that we should appreciate what Mother Nature

> left behind [for] us. ... My land is sitting over there crying. ...We see the scars the company left behind. Why should we let these companies come in and demolish everything ... in sight? They promise a lot of stuff, but right now I don't think I'd believe any of them anymore. I'm scared about the future for my grandkids. What will there be left for them? (Donna, activist).

An Hispanic woman similarly explained how she became politically active for the first time in her life after falling seriously ill and being fired from her job at a high-tech manufacturing plant in Albuquerque:

> I was an assembler ... and gradually got upgraded in different jobs, which exposed me more and more to the chemicals. ...The parts that we assembled were washed in freon, trichlorethylene 1-1-1, and some other solvent. They never gave us the name of it. ...We used to stick our hands in to wash these things. And after a few hours, after they dried up, well your hands were white, white, white. ... And if you smelled strong odors, they would say "It's nothing, just keep on working." You could see a film of white clouds [hanging over the work area] and we breathed all that stuff from eight hours to 10 hours a day. I worked for five years there. I was in the hospital because of my liver when they fired me. They have taken so much away from me and to end up with nothing. I know that no amount of money they could give me would ever bring me back my health. But maybe I could live my life a little more comfortable, you know? (Monica, activist).

Monica's experience motivated her to co-found an organization with the express purpose of seeking redress in the form of resources and medical services for other women poisoned by their jobs in the computer chip industry. The awards she and others received through litigation, however, barely covered their medical expenses. As a result, the organization has continued to lobby legislators for toxic victim compensation and stricter workplace safety laws and has actively worked to raise community awareness.

For Monica, inequalities based on gender, race/ethnicity, and class became visible as a result of her own experience; the vast majority of the assembly workers in the industry, and, therefore, those most affected by dangerous working conditions, were working class women of color like herself.

> People don't realize these companies are dangerous to work [for]. And you know, we don't want these companies to go away. New Mexico needs work. But these companies bring all the [top] people with them and we get all the shitty jobs. And that's not right. "We're bringing 500

> jobs to New Mexico." Well, big deal. We get all the dirty work (Monica, activist).

Another Hispana was similarly mobilized by the environmental damage to her barrio resulting from the lack of regulation of polluting industries and a nearby military base. Dalia was especially offended by the patronizing attitude of government consultants, whom she perceived as both classist and racist, and who failed to show respect for the wisdom of the existing barrio leadership:

> ...I got pissed. I came to a meeting and there were all of these people from the outside with their own agenda telling the poor people of color what was best for them. And they didn't even know it. And it got me very angry. I have a real hard time with people acting arrogant on soil that is foreign to them (Dalia, activist).

For these and other women in the study, race, ethnicity, class, and gender often surfaced as important considerations in their political struggles. However, the findings indicate that rage and personal experience are perhaps overplayed in the literature as causes of women's politicization. Seen here in this study, they are outflanked by other, more compelling motivations.

Tied closely to leaders' interest in issues of community health and safety and to preservation of the natural environment was concern for the well-being of the next generation. Concerns for public health and the environment were cited by 92 percent of both the Indian and Hispanic leaders as motives for their political participation. Again, for several of the leaders, spirituality was interwoven with their environmental beliefs. One leader echoed the perspective of other women when she explained:

> The earth is our mother. The earth is a very important female entity to me, as an Indian woman and as a Pueblo woman raised with very strong traditional beliefs. We were raised to have a lot of respect for those spirits or those entities that give us nourishment and life. And it's just a part of my own personal spiritual beliefs. And so it becomes an obligation for me personally, to do whatever I can as an individual out of respect and love for those spirits and those entities. To do whatever I can to protect it. Not only for the love and respect I have for it, but also because it's important for our children and grandchildren to have that same opportunity to use the earth and have it give like it's meant to be (Valerie, public official).

Maintaining environmental quality and conserving natural resources were themes that emerged frequently among the leaders, and which were directly related to concerns over the future of children today.

Representing children's issues, protecting their rights, and in particular, advocating for girls' empowerment, were motives for participation articulated by a majority of both the Indian (69 percent) and Hispanic (75 percent) leaders. One woman who has struggled for the right as a woman to participate in the governance of her tribe went so far as to say, "I don't care if I win or not as an elected official. My purpose is to let younger women know they can get on the council, they can be involved in their tribal government" (Maureen, public official). The leaders' advocacy for children and youth were often tied to their support of a women's agenda. More than half of the Hispanas (63 percent) and one-third of the Indian leaders (35 percent) specifically cited the advancement of women and women's issues as a motive for their own political involvement. All of the women interviewed, in one way or another, articulated their support for egalitarianism and the empowerment of women. We will explore the importance and influence of gender identity in greater depth in the next chapter.

As they matured and gained valuable leadership experience, these women's motives for political action often evolved through time, as illustrated in the following narrative:

> I thought when I was 15, I was gonna be a politician. ...I was running campaigns by the time I was 16. I thought I could make a difference; make it a better place for people to be in. And then I got to graduate school... [and] it really hit me how much people—to get into politics and to survive in politics—how much people coopt their own values. ... I knew I wasn't interested in that. [Later,] I met up with an old guy from the Teamsters who...taught me how to organize, how to negotiate. He taught me about a lot of stuff. I worked at that point with Latina women, Asian women—Japanese, Korean—working in fish houses; packing plants down in the worst part of Los Angeles. And that was the first time I had a taste of trying to make a difference. And I realized at that point that even though I didn't get involved in [electoral] politics, I could help level the playing field somehow (Miranda, activist).

Miranda's account also demonstrates the ambivalence felt by some of the leaders toward formal, electoral politics. She was not the only activist to express concern over becoming "coopted" by participating in institutional party politics. Indeed, several organizers rejected such involvement as "selling out." However, while some activists had no desire to enter the realm of institutional politics, preferring instead their community-based organizing activities, others used their experience as a stepping stone to public office. Yet other leaders in this study chose careers in politics or sought public office at the urging of friends and supporters.

LEADERS' PERCEPTIONS OF POLITICS AND POLITICIANS

> Politics has become such a dirty word, you hate to say you're a politician. A true politician represents the people, her constituents. The reservation boundaries are coming down fast. The more we know, the more we have a chance of holding on to what we have. I represent all women in the state. Not just Native American women. By not voting, they give up their power to affect change. Some people are afraid of losing their Indian identity by getting involved. They are more at risk of losing their identity by not getting involved (Daphne, public official).

The majority of the New Mexico leaders (80 percent) were members of the Democratic party, as we can see in Table 14, below. Specifically, 88 percent of the Hispanas (21) and 73 percent or 19 of the native women were registered Democrats. Only one Hispanic and two Indian women were registered Republicans.[16] While 73 percent or 19 of the activists belonged to one of the two major parties, all 24 of the elected and appointed officials were party members—despite the fact that the major U.S. parties do not play a role in tribal politics. Strong partisanship was not evident among this group of women as several said they voted a split ticket, depending on the issues at stake and the candidates running. In the case of the public officials interviewed, their willingness to cross party lines may also reflect the non-partisan nature of the offices that the majority held (see Table 4) in municipal or tribal government. Furthermore, several Indian leaders who held elected or appointed positions in state or local government felt that they had gained their seat due to their own efforts and qualifications rather than because of their party affiliation.

Table 14. Leaders' Political Party Affiliation

	Activists		Officials			
	Indig.	Hisp.	Indig.	Hisp.	Total[1]	%[1]
Democratic Party	7	11	12	10	40	80
Republican Party	1	0	1	1	3	6
Independent	2	2	0	0	4	8
Alternative Party[2]	3	4	0	0	7	14
Rejects Parties[3]	2	0	0	0	2	4

[1] When summed, the total column is greater than 50, and the percentage column is greater than 100 percent, because some of the leaders registered as Democrats or Independents also supported an alternative political party.
[2] Alternative Party: These leaders actively supported an alternative political party such as the Green or Labor parties but may also have been registered as a Democrat or Republican.
[3] Rejects Parties: In protest of the dominant political system, these leaders consciously chose not to participate in the electoral politics of the United States. They may, however, participate actively in the electoral politics of their native governments.

Grassroots activists, whether Indian or Hispanic, were less involved with either of the major parties. Of the activists, four (15 percent) had registered as Independents, seven (27 percent) were active in alternative parties such as the Green or Labor parties (six were simultaneously registered as Democrats), and two (eight percent) had withdrawn completely from formal electoral politics in protest of the existing system.

Most of the leaders interviewed equated political participation with public service, civic responsibility, and empowering others rather than with electoral or party politics. As one woman stated "We're all political. Even our great grandmothers were political. They were making sure they were providing for their families. It's all politics" (Dalia, activist). Many leaders felt that politics provides a way of contributing to the wellbeing of the community and defined politics as "decision-making with power." Moreover, both activists and public officials saw politics as an integral component of human nature. This perspective is reflected in numerous comments such as: "Politics is a way of life. You deal with it on a daily basis. There's politics in everything you do" (Tracie, public official). "Human interaction is political" (Lorna, activist). And, "I don't know how you distinguish politics from life" (Tammy, public official).

However, not all of the women viewed politics and politicians in a favorable light; some emphatically did not. For example, several leaders expressed the opinion that "politics [is] a dirty word." Valerie, a public official, noted, "I'm very political although I don't like that word—it has so many negative connotations. I see my role as having been grassroots that graduated to tribal leadership." Scholarship indicates that women in public office, in contrast to their male counterparts, generally view themselves as "public servants" or delegates of the people rather than as "politicians," and that they frequently devote more time to constituency work (Antolini 1984. Also see Richardson and Freeman 1995; Rosenthal 1995a; Darcy et al. 1987).

Several of the New Mexico leaders were careful to distinguish politicians from activists: "Politicians lie a lot. Activists have a vested interest in the community" (Rachel, public official). And, "Activists bring social consciousness to an issue. A politician wants something for themselves" (Tanya, public official). To a large extent, these sentiments mirror those of the public. Currently, the term "politician" carries such negative connotations in popular culture that many people consider it pejorative. Interestingly, many of the leaders articulating such opinions were themselves "politicians" serving in the role of elected or appointed public officials. As they had, in many cases, entered electoral politics from grassroots organizing, they clearly did not view themselves as "insiders" and strongly criticized establishment politics.

While some of the study participants grudgingly admitted to being political, neither the public officials nor the grassroots organizers readily identified themselves as "politicians." Instead, public officials such as Tanya and Rachel saw themselves as activists with a duty to reform the existing system making politics more honest, accessible, and responsive to Indian people. From this perspective, activists represent the concerns of their community, whereas politicians are out for their personal gain and individual gratification. Not surprisingly, several public officials who started out their political careers as community-based organizers also viewed favorably the participation of activists in the political process. Jacinta, an activist-turned-public official, noted from her own experience that, "Politics needs to have activists, or you get too complacent in government." Another leader expressed her belief in the centrality of grassroots activism to the maintenance of democratic politics.

> [Activism] is a really important profession in this country that is very undervalued. You can't have democracy if you don't have community, and communities don't happen naturally. Communities need organization, facilitation, resourcing to be able to adjust and grow and meet oncoming challenges and opportunities. There's no way to have a democratic society if you don't have community organizers (Marta, activist).

Despite the overwhelming evidence of their leadership, several women were reluctant to acknowledge their considerable political power. Not only did they seem surprised when referred to as "leaders," but some showed disbelief when told that they were perceived as powerful in the political arena. This finding is remarkably similar to Carol Hardy-Fanta's in her 1993 study of Latina leaders in Boston politics. As with Hardy-Fanta's sample, the New Mexican women I interviewed defined themselves, their lives, and their work in terms of their relationship to their families and community. As Ackelsberg writes,

> For women, at least, the community—constituted of networks of friends and neighbors—*is* one locus of the development of...political consciousness. It is in and through such networks, located at the interface of personal/household concerns and the impact of employer or public policy decisions, that many...women engage in collaborative activity and begin to experience themselves, and others, as...competent social-political actors (1984, 257. Emphasis in original).

In both Hardy-Fanta's and my findings, the role of these leaders as mother, wife, woman of color, professional, or citizen superceded their identification as politically influential members of their community. This is because, as Ackelsberg explains, "...women live in a context of webs of relationships, informal and formal, which structure their patterns of interaction, give meaning and order to their lives, and may well be crucial factors affecting the development of their political consciousness" (1984, 251). This finding, which appears to hold true for women regardless of their race or ethnicity, has been reported time and again in recent scholarship (Krauss 1998; Hoikkala 1995; Gilkes 1994; Bystydzienski 1992a; Di Chiro 1992; Fowlkes 1992; Ford 1990; Pardo 1990; West and Blumberg 1990; B.A. Carroll 1989; S.J. Carroll 1989a; Kingsolver 1989; Morgen and Bookman 1988; Susser 1988; Molyneaux 1985; Kaplan 1982). This may be due, in part, to women's gendered socialization. As Fowlkes explains,

> ...*Any* woman who has "become political," whatever her race or economic class or sexuality, has had to negotiate two sex/gender barriers: one to consciousness that she herself can act politically and the other to recognition by others that she can act and does act politically (1992, 96. Emphasis in original).

On the other hand, an activist-turned-public official conceded, "Politicians seek power. [But] it's not a negative—someone has to be willing to make the hard decisions" (Magdalena). Similarly, a few leaders acknowledged their power or sought more, noting the drawbacks to being either an activist or a public official. For example, a local school board member who was considering running for mayor did not see herself as a politician, or as powerful, even though she regularly participated in policymaking and advocated for native students and their parents in the school district. Flora felt that she could accomplish more for Indian youth in her community as mayor. However, she had doubts about her chances of winning a municipal election. "There is more power as a politician...but I'm not sure the community is ready for a Native American woman's leadership." And, apparently, neither was her pueblo. Flora, like several of the Indian women interviewed, was limited to participating in politics "outside," as her pueblo prohibited women's involvement in tribal government.[17] We now turn to a discussion of these women's diverse paths to leadership.

ACQUIRING POWER: PATHS TO POLITICAL LEADERSHIP

> When the money came in from the state of New Mexico to establish this program, to work at the grassroots level...they wanted to employ someone from the community. Then this opportunity came up. I'm from the community and I have the [qualifications]. And I know who gets along and who doesn't. If you don't get along, it doesn't matter how much education you have. You have to live this work. It has to be your life. This is my life (Dalia, activist).

Grassroots organizing has long been a political practice vital to the development and maintenance of democracy. As an expression of dissent, or as the voice of marginalized groups' interests, it has provided an essential outlet for those excluded from public decision-making processes. Historically, women have effectively utilized community-based organizing as a forum for making demands on political and economic institutions, and for seeking social change. Besides providing women with valuable skills, confidence, and leadership experience, the exposure afforded in such activism both allows women to "test the waters" if they are considering entering formal politics, and gets their name out in front of voters and party elites. Community activism, then, has often served as the springboard to women's attaining positions of power in formal, institutionalized politics. Fourteen or 58 percent of the 24 New Mexico officials interviewed entered electoral politics following their involvement in grassroots organizations.

While the New Mexico leaders generally shared similar motives for entering into political life—to acquire and exercise influence and thereby effect change for their communities—their paths to leadership varied substantially. Ninety percent, or 45 of the 50 women were involved in either policy-making relative to environmental justice or in general advocacy for communities of color. Of the 26 activists, 22 women (85 percent) chose to work toward their public policy goals through non-governmental organizations (NGOs) pursuing environmental justice. Of the four remaining activists, one volunteered her environmental law expertise to her tribe rather than to an NGO, one dedicated her volunteer efforts to ending violence against women and children, and two worked primarily for tribal sovereignty rather than for environmental justice. Of the 26 grassroots activists, 23 (88 percent) were directly tied to one or more established groups either as volunteers or paid staff. (See Table 1). Three of the 13 volunteers were "independent" activists who were not formal members of any single organization, although they expended considerable time and resources as volunteers for a number of different organizations. In general, the grassroots groups existed outside of, but interacted with, formal political institutions such as legislative bodies, the courts, and executive agencies.

The volunteers exercised policy-making authority by serving on their organizations' board of directors or steering committee. Similarly, staff members shared in developing organizational goals, identifying policy and program directives, setting budget priorities, and formulating issue campaigns. Three of the volunteers and seven of the staff participated in founding their particular organization. With the exception of the "independents," all of the activists sought to effect changes in public policy as leaders within their grassroots groups. The independent activists also exercised leadership although their situations were somewhat different. For example, Donna, who was a vocal advocate of tribal sovereignty and worked with several groups to change environmental policy, was establishing an organization for environmental justice in her pueblo when I interviewed her. Carmen, a semi-retired native rights leader, remained politically active by providing technical assistance and valuable organizing experience to fledgling environmental groups in New Mexico, Arizona, and Nevada. Karen's knowledge of environmental law, her advisory role to the tribal governor, and her advocacy on behalf of her tribe during critical negotiations placed her in a leadership position as an environmental activist for her tribe. In varying ways, these activists influenced public policy within both Indian tribes and the state of New Mexico.

Although there is some overlap, the paths to leadership taken by the 50 women interviewed can be traced to three general sources. First, women mobilized as youths participated in social justice causes while in high school and/or college. Their exposure to varied political ideologies and their involvement in social movements and/or party organizations, provided them with contacts and valuable experience. In many cases, this experience helped the young women to determine their future career goals and life choices, cementing their commitment to political activism. Second, while fulfilling their roles as mother and homemaker, several women developed valuable planning and organizational skills, which aided their ascent to leadership. Women active in their children's education found occasions to influence school policy and programs. In numerous cases, their experiences led to their serving on local or state school boards, and these positions led to yet other opportunities for political involvement. Similarly, women free to volunteer their time initiated or joined efforts to protect, maintain, or enhance the quality of life in their communities. As spokespersons for their neighborhood associations, for example, women gained exposure and experience in dealing with bureaucracies, local officials, and the media. As they gained prominence, these women were invited to serve on city boards and commissions. Finally, women who volunteered their time to work for a political party and/or service organization gained experience, useful insights to political processes, administrative skills, and contacts. In some cases, women benefited from mentorship, specific leadership training and development opportunities, and financial or other support when running their own campaigns for public office.[18]

The New Mexico leaders were a diverse group with regard to their demographic characteristics, their social class as expressed in terms of their "insider" or "outsider" political status, and their trajectories into politics. However, despite these differences, they shared common goals and many reported comparable experiences of political socialization. With few exceptions, these women were not catapulted into politics by some life-altering catalytic event. On the contrary, most were socialized into public life early on by family members and others who helped to prepare them for assuming positions of influence in their communities. Most significantly, regardless of their political status as "insider" or "outsider," these women expressed similar motives for assuming leadership: the desire to fulfill their civic obligation by improving the quality of life for residents of their racial/ethnic communities. We now turn to chapter 5 and a more extensive examination of the role that racial/ethnic identity and gender play in the politics of this group of leaders.

CHAPTER 5
The Role of Identity and Political Ideology

> My mother told me as I was growing up: "You're a woman and you're a Mexican and you just have to work harder and that's the way it's going to be" (Jacinta, public official).

> Issues of race, class, cultural identity, and gender subordination intersect and coexist. ...The multiple levels of consciousness that emerge from different contexts of oppression within U.S. society ... contribute to structuring identity: a collective national/cultural/racial consciousness, a feminist/feminine consciousness, and a class consciousness (Acosta-Belén 1993, 182).

MUCH OF AMERICAN WOMEN'S POLITICAL INVOLVEMENT HAS BEEN traditionally motivated by issues pertaining to race, sex, class, and economic concerns (Gilkes 1994; Cable 1992; Tolleson Rinehart 1992; Fincher and McQuillen 1989; Susser 1988; Young 1988; Dabrowski 1987; Ackelsberg 1984; Githens 1983). Indeed, race and ethnicity, together with sex and class, are at the very center of America's political and social history. Race has been cited as a major force in the creation of mass movements (McClain and Stewart 1995; Omi and Winant 1986; Morris 1984), in the formation of state policy (Reed 1988; Blauner 1987; Takaki 1987; Carmines and Stimson 1986), in the individual's political socialization (Jarvenpa 1989; Bennett and Bennett 1986; Jankowski 1986; Parenti 1967) and in their political participation (Horton 1992; Garcia et al. 1991; MacManus and Cassel 1988). Women of color who bear the double burden of sex- and racial discrimination are often economically and politically marginalized as well (Liebowitz and Carroll 1996; Sen 1995; Baca Zinn and Thornton Dill 1994; Jaimes 1992; LaDuke 1991; Albrecht and Brewer 1990; King 1987; Segura 1987; Thornton Dill

1987; Gurin 1985; hooks 1981; Cotera 1980). Consequently, their activism is influenced not only by their experiences of sexism, but also by the effects of oppression and racism on their lives, on the lives of their families, and on their communities. Not surprisingly, a growing body of scholarship reveals that both racial/ethnic and gender identity inform and influence the political participation and policy preferences of women of color. As Cohen, Jones, and Tronto observe,

> [S]ystemic forms of domination operate to shape the lives of women and men in our society and to create sites of power/resistance. ...[R]acism, sexism, class exploitation, homophobia, and other institutions of control are interrelated sites of domination and struggle that have left a long legacy in U.S. politics... These systems of power create political divisions such as race, gender, economic and social class, sexual orientation, ability, ethnicity, religion, age, region, and language by making these categories formative of our sense of who we are and where we stand. Furthermore, these categories constitute identities whose complexity we cannot grasp when we isolate discourses of race, class, gender, or sexuality from one another (1997, 3).

Racial/ethnic- and gender identity both informed the political ideology and influenced the policy agendas of the majority of the New Mexico leaders interviewed for this study. We now explore in greater detail the role that race, ethnicity, and gender played in motivating this group of leaders to enter public life, in shaping their ideologies—in particular, their conceptualizations of "feminism" and "environmentalism," and in formulating their political goals.

POLITICS AND THE INTERSECTION OF RACE/ETHNICITY AND GENDER

> It hasn't been easy. I've got three [strikes] against me: my looks, the fact that I'm a woman, and that I'm Hispanic. I've had to constantly prove myself throughout my whole career (Victoria, public official).

Culture and race/ethnicity were highly salient for the New Mexico leaders interviewed. Many women felt strongly about the use of racial/ethnic categories. How they identified themselves was a consciously political decision.[1] For example, Hispanas who considered themselves to belong to peoples indigenous to the region proudly acknowledged their mixed heritage by calling themselves *Chicanas* or *Mestizas*. On the other hand, one Hispana who claimed "pure" Spanish heritage was offended when I asked whether she considered herself Chicana or Mexican American. She insisted that she was caucasian and a descendant of the *conquistadores*, presumably implying that she had no Indian blood despite her family's presence in the region for over 400 years.

In contrast, several of the Indian leaders and Hispana activists considered themselves to be members of colonized groups who continue to suffer racial, social, and economic oppression at the hands of the dominant Anglo American culture. A smaller group of women expressed militant or nationalistic views, which underlined their concerns with race, class, and sovereignty. They identified as "indigenous," "third world," or "Chicana." However, while terms such as "Chicana" and "Native American" are used with frequency in academic circles, few of the leaders whom I interviewed actually used these terms themselves. Instead, the great majority of American Indian women identified as members of a particular tribe and used the generic term "Indian" to refer to other Native American peoples. Women of Spanish or Mexican heritage described themselves as Mexican American or Hispanic.

Several women used racial and/or ethnic descriptors to self identify, often in combination with their gendered roles. For example, these were some of their responses to the question "How would you describe yourself?"

- A Chicana working mother; a grandmother; a strong woman.
- An independent Comanche woman; an Indian activist.
- A Native American woman functioning in a contemporary setting...yet trying to maintain traditions.
- A Hispanic woman professional; a single mother.
- An Indian person who strongly believes in Indian spiritual ways.
- A Chicana feminist.
- Not your typical Pueblo woman [laughs heartily].
- A brown ethnic woman.
- A person caught between two worlds.

The leaders' responses revealed a strong identification with their racial/ethnic group and culture. In particular, four aspects or indicators of racial/ethnic identification emerged from patterns in the data – self-labeling by race/ethnicity, racial/ethnic consciousness, racial/ethnic salience, and cultural motivation. While four different indicators are cited, they all refer to the same phenomenon – racial/ethnic identification.

Self-labeling by race/ethnicity refers to leaders who self-identified using racial or ethnic labels such as "American Indian," "Chicana," "Mexican American." Racial/ethnic consciousness, on the other hand, is an adaptation of Sue Tolleson Rinehart's notion of "gender consciousness," discussed below. In this case, New Mexico leaders who identified with other members of their racial/ethnic community *as a group*, displayed a positive affect toward members of their racial/ethnicy group, and/or demonstrated a sensitivity to the group's sociopolitical situation and well-being exhibited racial/ethnic consciousness. Similarly, leaders who used their political activism to address racism and/or issues of race, ethnicity, and culture demonstrated racial/ethnic salience. Finally, leaders who were motivated to enter politics by a desire to preserve their culture exhibited cultural motivation.

Table 15. Factors Indicating Leaders' Racial/Ethnic Identity

	Types of Leaders				
	Indig.	Hisp.	Activists	Officials	N=50[1]
Racial Self-Labeling	8(62%)	7 (54%)	2 (15%)	3 (27%)	20 (40%)
Racial Consciousness	8 (62%)	6 (46%)	5 (38%)	4 (36%)	23 (46%)
Racial Salience	13 (100%)	11 (85%)	12 (92%)	4 (36%)	40 (80%)
Cultural Motivation	13 (100%)	12 (92%)	12 (92%)	7 (64%)	44 (88%)

These four categories were not mutually exclusive; rather, they tended to overlap. Furthermore, as we can see from Table 15, above, the New Mexico leaders frequently displayed two or more of these indicators of racial/ethnic identification. For example, an indigenous leader might describe herself as an Indian activist (racial self-labeling) working to empower her people (racial consciousness) by championing tribal sovereignty (racial consciousness and racial salience) in order to preserve their native language and culture (racial consciousness and cultural motivation).

Of the 20 women who described themselves by their racial/ethnic group, 10 were indigenous (38 percent) and 10 were Hispanic (42 percent), 15 were grassroots activists (58 percent) and five were public officials (21 percent).[2] Nearly half of the New Mexico leaders (23 or 46 percent) expressed racial/ethnic consciousness by identifying with other members of their racial/ethnic group *as a group*, displaying a positive affect toward group members, and a sensitivity to their sociopolitical situation and well-being.[3] In particular, 13 native women (50 percent), 10 Hispanas (42 percent), 14 activists (54 percent), and nine officials (38 percent) exhibited racial/ethnic consciousness.

Furthermore, a large number of leaders (40 or 80 percent) demonstrated the salience of race/ethnicity to their practice of politics. In this case, all but one of the indigenous women (25 or 96 percent), 15 of the Hispanas (63 percent), 24 activists (92 percent), and 16 officials (67 percent) employed their political activism to address specifically the problem of racism and other issues of race, ethnicity, and culture. These findings are consistent with the existing literature on Hispanas' and native women's activism. For example, in her study of public officials in California, Cruz

Takash reported that Latina leaders were generally more concerned with "issues facing the Latino community as a whole such as employment, access to education and retention, and safe neighborhoods, issues largely stemming from institutionalized racism and classism" than they were with promoting a feminist agenda of concerns specific to women (1993, 354). Other studies have similarly noted that American Indian women's activism is often specifically directed at improving conditions for their native communities (see for example Hoikkala 1995; Jaimes 1992; Ford 1990; Albers 1989 and 1983). As one researcher notes,

> politics in most contemporary Indian communities are centered around basic reproductive issues—improving jobs, housing, education, and health care as well as protecting community rights in land, water, and mineral resources that can provide for an improved standard of living (Albers 1989, 147).

Finally, as noted in chapter 4, cultural preservation provided an incentive for entering politics for nearly all of the native leaders and grassroots activists (25 or 96 percent) and for 19 (79 percent) of the Hispanas and public officials. As Pueblo author Paula Gunn Allen has noted,

> To survive culturally, American Indian women must often fight the United States government, the tribal governments, women and men of their tribe or their urban community who are virulently misogynist or who are threatened by attempts to change the images foisted on us over the centuries by whites. The colonizers' revisions of our lives, values, and histories have devastated us at the most critical level of all—that of our own minds, our own sense of who we are (1986, 193).

If we simply examine the frequencies for each of the four groups of leaders relative to the four indicators of racial/ethnic identity, it appears that identification with their racial/ethnic group was generally more important for the indigenous leaders than it was for the Hispanas, as well as more important for the grassroots activists than for the public officials. As one native leader stated,

> The first thing I think about myself is that I'm an Indian; I don't think that I'm a woman. I'm an Indian *person*. ...In the Mohawk culture, women are pretty strong anyway. Pretty influential... It's more important for me to be a Mohawk than to be a woman (Kari, activist).

The salience of racial/ethnic identity is grounded in the day-to-day experiences of the women interviewed. While New Mexico is often represented as a culturally diverse state that enjoys harmonious racial relations, several leaders told a different story as noted in the quotes by Victoria and Jacinta earlier in this chapter.[4] Camila, a public official, gave a forceful assessment of New Mexico's political and economic environment from the perspective of Indian people.

> Right now Native Americans in the state are...just now making very small inroads to the legislature. And that's just been in the last 10 to 15 years. That's nothing! ...That's lamentable for New Mexico. [Voice rising and becoming very animated] You know, so long as we're just dancing and making our pottery, that's okay. "But you just stay over there!" [gesturing emphatically]. Tourists—"Ya, we can come in and make money off of you but [Indian people] can't have any part of it. Just keep in your own place!" That's the message that comes out. [Pause] Just being Native American, there's big mistrust. There's a lot of misunderstandings about where Native Americans fit... "They are not part of our political fabric at all." And so people feel very uncomfortable about that. "Oh, so I guess you have a pretty good life. You get a monthly check from the government and a new truck every year!" To this day, even educated people [believe this]. They think we don't pay taxes. There are huge misconceptions. That contributes to this mistrust and somehow we are not really participants; we are some other species [stated with indignation].

Camila's narrative illustrates passionately the ways in which racial oppression, as Gilkes says, "combines cultural humiliation and destruction, political subordination, and economic exploitation to maintain a hierarchy that limits the life chances of a group of people. ...[It] places the entire community in a colonial relationship, a relationship of powerlessness and dependency, within a dominant and dominating society" (1994, 232).

The anger generated in response to racism often fueled the New Mexico leaders' determination to enter into politics or to continue with their community work. As one Indian official noted, "Anytime you have a woman of color, color comes out more than the male/female issues" (Camila). Their experiences of racism and of sexism were often entwined, however, as is evident from Magdalena's narrative.[5]

> I'm a very realistic woman, I knew that our city was not ready for a woman to be elected mayor, particularly five foot tall and five foot wide, a grandmother, Hispanic; there was no way I would be taken seriously. But, because when you run for Mayor you go to approximately 50 forums...I would have a perfect opportunity to appear at their sponsored forums and I could get my word said. So I decided I would run for mayor. And it was probably one of the best times in my life because I had no illusions. But I had an agenda. And it was working very well for me. ...I like people, I like to talk. I was very versed on

> all the issues in the city so I was able to come across very well and hold my own very well. After the forums women would come up to me and you know they would be surprised. "Oh _____, I didn't realize" and "Oh, you're so smart, good job." And then I would say "Would you vote for me?" And you could see them take a step back. ...And some of them would say "Oh _____, if you would run for county clerk, if you would run for secretary of state, you'd win."[6] And I said, "But I don't want to be county clerk, I don't want to be secretary of state, I want to be the Mayor." I think they were kind of shocked at my audacity to even consider putting myself forward even though ever since we've had the mayor/council form of government, there's always been a woman to run (Magdalena, public official).

Knowing that she had only a very slim chance of winning the election, Magdalena ran anyway, as a strategy for making her ideas heard and for influencing change in the community. Gilkes similarly noted in her study of African American women's community work that the prevalence of racism in social institutions motivated black women's activism to battle racism and "empower their communities to survive, grow, and advance in a hostile society" (1994, 230).

Like the African American women in Gilkes' study, the New Mexico leaders found positive ways to combat racism in their communities by building social networks among people and by strengthening local institutions. Both indigenous and Hispanic leaders were active in developing programs that celebrated heritage through language, dance, storytelling, art, and crafts in order to preserve their communities' cultural traditions. Leaders also participated in cultural exchange programs to better develop cooperative relationships with other racial/ethnic groups in the state. This finding is consistent with the literature because as Ackelsberg (1984) and others have found, women's collective action to improve the quality of life in communities is often built upon existing networks and serves to unite people across racial and ethnic lines (see for example Baca Zinn and Thornton Dill 1994; Albrecht and Brewer 1990; Morgen and Bookman 1988; Dabrowski 1983).

For many women of color, the role that race/ethnicity and gender play in forming a woman's identity are inseparable, and somewhat complex. As we saw earlier, several leaders described themselves in terms of their multiple identities, as a "Chicana working mother, grandmother, [and] strong woman" in one case, and as an "independent Comanche woman [and] Indian activist" in another. In general, the New Mexico leaders held a multiplicity of sometimes conflicting roles and responsibilities that resulted from their being women on the one hand, and from living in two different worlds—their own community and the dominant Anglo American society – on the other hand. Frequently, women expressed feeling torn between cultural gender expectations and their individual needs as women, mothers, workers, and leaders. The conflict arising from these competing demands is particularly evident in the following leader's narrative,

> ...because I'm an Indian woman, there are certain roles I have to play that may be seen as subservient by some. But that's what you have to do. That's tradition. With tradition comes a certain amount of resistance to change and that's bad. But if you move too fast you lose your culture and that's bad. ...I'm involved with women's issues...I'm pro-choice. Times are changing. Yes, we're Pueblo women but [men] have to help. It's a two-way process. We go to work, have a career and are expected to be subservient, cook, clean, and take care of the children. The Indian population, especially the Pueblo population, is very traditional, matriarchal in some sense, but politics are very male dominated. Now women can attend [tribal] council meetings and vote in [my Pueblo] but not when my mom was younger (Tanya, public official).

As with the four indicators of their racial/ethnic identification, clear expressions of the leaders' gender identity also emerged from the data, which I grouped into four categories – self-labeling by gendered roles and/or sex, gender consciousness, gender salience, and gender motivation. Although four factors are cited, they are all representative of the same phenomenon—gender identity.

Gender self-labeling occurred when leaders self-identified by their specific sex and/or gendered roles such as "woman," "mother," "wife." As stated above, the term "gender consciousness" is borrowed from Tolleson Rinehart (1992) and refers to leaders who identified with other women as a group, displayed a positive affect toward women, and/or demonstrated a sensitivity to women's sociopolitical situation and well-being. Gender salience was exhibited by leaders who used their political activism to address women's issues by advocating for and/or empowering women. Lastly, leaders expressing gender motivation admitted to becoming involved in politics in order to advance the rights of women, to promote women's issues, and/or improve the status of women.

Again, as we can see from Table 16, these four overlapping categories were not mutually exclusive. Both indigenous and Hispanic leaders frequently exhibited two or more of these indicators of gender identification. To illustrate, an Hispanic legislator might describe herself as a woman and grandmother (gender self-labeling) working to empower single mothers (gender consciousness and gender motivation) by sponsoring legislation to benefit women and girls (gender salience and gender consciousness) by providing them with job training and subsidized child care (gender salience, gender consciousness, and gender motivation).

Table 16. Factors Indicating Leaders' Gender Identity

	Grassroots Activists		Public Officials		
	Indig.	Hisp.	Indig.	Hisp.	Total/%
Gender Self-Labeling	8 (62%)	6 (46%)	5 (38%)	4 (36%)	23 (46%)
Gender Consciousness	9 (69)%	13 (100%)	11 (85%)	11 (100%)	44 (88%)
Gender Salience	1 (8%)	8 (62%)	6 (46%)	6 (55%)	21 (42%)
Gender Motivation	2 (15%)	7 (54%)	7 (54%)	8 (73%)	24 (48%)

Several leaders (23 or 46 percent) described themselves in relation to their sex as, among other things, a strong woman, single parent, working mother, wife, grandmother, and "supermom." Thirteen of these leaders were native women (50 percent) and 10 (42 percent) were Hispanas; 14 were grassroots activists (54 percent) and nine were public officials (38 percent). A majority of the leaders (44 or 88 percent) demonstrated gender consciousness. Similar to other forms of group consciousness, it represents an identification with others like one's self, a positive affect toward them, and a sensitivity or connectedness with the group and its well-being (Tolleson Rinehart 1992). Most of the public officials (22 or 92 percent) and grassroots activists (22 or 85 percent), over three-quarters of the indigenous leaders (20 or 77 percent), and all 24 of the Hispanas identified with other women *as a group*, displaying gender consciousness.

As we can see from Table 16, gender was especially salient for a slightly smaller number of leaders (21 or 42 percent), but more so for the public officials (12 or 50 percent) than for the grassroots activists (9 or 35 percent). Further, seven native women (27 percent) and twice as many Hispanas (14 or 58 percent) addressed women's issues specifically through their political activism by advocating for and/or empowering women.

> I'd like to talk more to women's groups to empower, [to explore] what solutions, changes can you make? To [give] a sense of empowerment for self and children's sense of pride in self, in their heritage (Toni, activist).

Finally, a larger percentage of the Hispanas (15 or 63 percent) than Indian leaders (9 or 35 percent), and a greater proportion of the officials (15 or 63 percent) than activists (9 or 35 percent) were motivated to participate in politics in order to advance the rights of women and to promote women's issues.

In their study of contemporary American women leaders, Le Veness and Sweeney (1987) described women's issues as public policy that disproportionately impacts women, including such concerns as abortion rights, funding for childcare, equal employment opportunities, job training for displaced homemakers, and economic security for elderly women (also see

Antolini 1984; Boneparth 1984; Sapiro 1981). These authors defined a feminist as

> a woman or man who wants these policy questions decided in a manner that gives women the maximum opportunity to participate fully in our society without discrimination. ...[And a] feminist politician is one (male or female) who devotes substantial career time to furthering women's issues (1987, 2).

Despite their advancing women's issues and fully believing that women should have the "opportunity to participate fully in our society without discrimination," numerous New Mexico leaders did not view themselves as feminists, however. (We will explore this paradox later on in this chapter.)

It is interesting to note that while the significance of racial/ethnic identity was generally greater than gender identity for both the indigenous and Hispanic women, racial/ethnic identity was more important for a larger percentage of indigenous than Hispanic leaders as well as more important for a larger percentage of the grassroots activists than public officials. At the same time, gender identity was more salient for a greater proportion of the Hispanas than native women, as well as more salient for a greater proportion of the public officials than the grassroots activists. A review of the frequencies for each of the four groups of leaders (Indigenous, Hispanic, Grassroots Activists, and Public Officials) relative to these indicators reveals that gender identity was generally more significant for the Hispanas and for the public officials than it was for the indigenous leaders and grassroots activists.

These findings are consistent with research on native women's activism, which generally tends to advance the well-being of the community as a group rather than the needs of Indian women in particular (see for example Hoikkala 1995; Jaimes 1992; Ford 1990). Ford explains that this perspective originates in the belief "that there is an interdependence, a complementarity, between [American Indian] men and women and that until their group has equality with other ethnic groups, they, as women, will not gain" (1990, 88). On the other hand, Chicanas have been urged to

> express their needs, not to strangers but to their own community. Like mothers liberating themselves from the often psychotic martyr complex [Maria complex or "marianismo"], Chicana feminists will have to shed their group martyrdom and no longer tolerate Chicanos who have for decades taken advantage of their accommodating behavior (Cotera 1980, 233).

A critical difference between the standpoints of the Chicana and native woman lies in the unique political status of Indian nations. Gender equity is less important an issue for indigenous women than tribal sovereignty, for instance, which equates with their very survival as a people and incorporates the preservation of native culture, lands, religious beliefs, and language. Gunn Allen's admonition illustrates the weightiness of this concern.

> We must strive to maintain tribal status; we must make certain that the tribes continue to be legally recognized entities, sovereign nations within the larger United States, and we must wage this struggle in many ways—political, educational, literary, artistic, individual and communal (1986, 193. See also Albers 1983).

While the New Mexico leaders, as a group, demonstrated greater racial/ethnic identity than gender identity, this is not to say that public policy relevant to women was unimportant to them. On the contrary, as Tanya's quote indicates above, they were very much aware of the social, economic, and political constraints faced by many women simply because of their sex. For example, one tribal leader sought office specifically to provide a voice for women's concerns.

> I see a need for tribal membership, trying to get back enrollment for women. In our community, if a woman marries outside, her children are not enrolled [in our Pueblo]. They are enrolled with the father's tribe. If the husband is non-Indian, the children are not enrolled anywhere. That has been an issue as long as I can remember. This year's tribal council has tried to address this, last year a committee was established. There's a lot of work that needs to be done. We need to look at the woman's side, trying to enroll her children. Also, the social problems need to be addressed. If any community is to survive, they need to address those problems of families and children. Those are the future leaders. Economic development is very important but if you start forgetting your children and your social problems, there's something wrong. You need to take care of your children, your people in the community (Maureen, public official).

Not surprisingly, gender issues were sometimes more salient for native women whose Pueblos prohibited or limited severely the participation of women in the political life of their tribe than they were for individuals such as Kari, who experienced few if any obstacles *as women*.[7] For example, contrast Kari's statement above, with that of Karen, a Pueblo woman.

> I'm a woman caught between two worlds. I'm a [feminist and] professional outside. But traditional and, I hate to use the word, submissive at home. For example, letting the men eat first, keeping your eyes down. I just do it [without thinking about it]. Mom taught us to be independent, self-reliant. We come from matrilineal religious societies.

> It was the Spanish influence that undermined this system by introducing patriarchy (Karen, activist).

However, as the following statements reveal, some leaders were ambivalent about introducing greater opportunities for women's political participation within their tribes.[8]

> We need to have a [tribal] administration that's sensitive to the community they serve. I don't know that it will change with more women serving. I hope our daughters will be involved. When you don't have women involved, it leads to oppression... (Toni, activist).

As another woman from a very traditional Pueblo noted,

> I'm not ready for a woman on the council. But I think I'm an influence. I can talk to different councilmen and let them know what I think and they can take it from there. ...I do know how to go into council and not be aggressive. ...If you're a woman here, it's not too smart to be really aggressive. They'll shut you down. If you work with [the councilmen] and present ideas like they're their ideas, and it's for the whole, it'll come across better. I go to the village meetings. There are some that don't allow women, but our village does. I go there, I give my opinion. I'm active there. [Pause]...It's not perfect, but it's okay. It will change. In the future, women will be more active in tribal politics. Women's clans lead now. When a child is born, it takes the mother's clan. The head of the house is the mother. The house is left to the daughters, the fields go to the sons. It's changing, but that's tradition. Before the Spanish came and the Americans, women were more involved (Rachel, public official).

The sense of feeling torn between tribal traditions and the desire for full and equitable participation in the civic life of her community emerges from Rachel's narrative.

Despite the political and social barriers imposed on women by their tribes, Indian women found opportunities to exercise leadership in the political arena within their tribe when possible, and outside of their tribe, when necessary. While women were prohibited from serving on (or even voting for) their tribal councils, several of the women interviewed provided technical assistance to their tribal governments in their professional roles as administrators in public agencies. Their position and expertise, however, did not guarantee that they were welcome in the male domain of tribal government. The irony of their situation was not lost on these women. The comments of an attorney in the Bureau of Indian Affairs illustrate the difficulty faced by women dealing with very traditional tribes.

> I have to go in to some of the very traditional tribal councils. As an example, there are some tribal councils that do not allow women in the room, and I've had to go in and make presentations. You can almost see it, they hate to listen to what I have to say. But once they listen, they realize that I know what I'm talking about. And I've had some very uncomfortable experiences where I've taken a male subordinate into a council meeting and they say, "Okay, Tom, give us your presentation" and they just completely blow me off. And he's like "Wait, this is my boss" and they say, "No, Tom, why don't you tell us." First of all, you can't believe that it's happening. But being raised in the traditional way, it's not surprising (Karen, activist).

In their capacity as elected or appointed officials on state or local boards or commissions, indigenous leaders represented their native constituencies and made use of opportunities to advocate for Indian women's concerns.

> I had to speak before the All Indian Pueblo Council only two months after being made [an organization] president. I said in my speech that women's issues are the same issues that men deal with as tribal issues. We deal with them as sisters, mothers, wives. We want to make sure that they listen to us. The education of children [for example] is still the wife and mother's role though we are not consulted on [the tribal leaders'] decisions. Maybe, hopefully, I started something. I gave them something to think about (Daphne, public official).

Women from tribes that maintained more egalitarian practices (such as the Comanche, Diné, Mohawk, Nambe, Pojoaque, and Western Shoshone), were more likely to enjoy a higher status within their tribe, have had female role models, and have opportunities for real political involvement in their tribal government. However, even women from such tribes expressed support, through their activism, for improving the lives of women in general. As one tribal leader stated,

> Women have been involved [in our Pueblo's politics] right from the start. Women have been serving on the council as long as I can remember. I can't speak for other Pueblos. I don't know why they don't involve their women because they certainly should. That's my opinion. They have a lot to contribute. For whatever reason, maybe they're more traditional. I really don't know, [but]...I'm a firm believer that every decision that is made affects every man, woman, and child living on the reservation. I think that we all need to have our input and be listened to. ...Even outside the Pueblo, a lot of women have come up to me and said "Gee, I admire you" for doing this, or doing that, or whatever. And they, in turn, are becoming more involved in their communities. I tell them, "You don't necessarily have to be a councilmember or an elected official. You can do things right at your own level, whether

> it be dealing with the senior center, or the day care centers, with housing. There should be things that come up that you can be effective in. Approach council-members and let them know that you're willing to help them write proposals, to get funding for different things that are needed in the Pueblo. Not necessarily being pushy because they won't accept that, but letting them know that you are sincere and want to help wherever you can" (Tracie, public official).

Women's involvement in the political arena was seen as important and necessary by the New Mexico leaders. Furthermore, there was a general perception that women's policy priorities are different from men's and that women can positively transform the way in which politics is played by making it more participatory and representative of marginalized group's interests. This was evidenced in statements such as: "Women, especially, should run for office. We bring a particular point of view and temperament to everything we do" (Estela, public official). Also, "Women's leadership is different from men's. As a group, we don't have the need to be top dog. [Instead] we ask: How can we both come out ahead?" (Sonia, public official). Similarly, when asked why she entered politics, one native leader responded, "I felt there was a need for more women to be involved in their tribal government...to be in that decision process. There's a lot of things that affects women and children and families that women are more in tune to than men" (Maureen, public official). These comments reflect the gender consciousness of the leaders and their sensitivity to the needs and perspectives of women as a group. We turn now to a discussion of the role of gender consciousness and feminism on the politics of the New Mexico leaders.

WOMEN, POLITICS, AND GENDER CONSCIOUSNESS

Maternal feminists, such as Jean Bethke Elshtain (1981), advance a politics of care characterized by traditionally feminine domestic concerns. They attribute women's "ethic of care" to the socialization of girls, which emphasizes human relationships, and to woman's unique experience of motherhood, which substantially influences women's identities and values and solidifies the ethic of care (Sidel 1995; Tronto 1993; Ruddick 1980). This perspective was evident in the words of one New Mexico leader whose activism was informed, in large part, by her gendered roles of woman and mother.

> Women are life-givers, their connection to Mother Earth is important. Women feel the impact of environmental issues more because they are in the home. Women look after the health and safety of the family and, by extension, the community (Terri, activist).

Proponents of maternal feminism argue that girls and boys are generally raised differently and that, as a result of their gendered upbringing, girls develop an ethic of care centering on relationships. In contrast to the communitarian orientation of girls, boys are socialized with an emphasis on individualism. Citing Carol Gilligan (1987), maternal feminists contend that women's ethical orientation is reflected in a web-like contextual ethics of care and reciprocal responsibility, which is based on "important differences in moral reasoning between men and women, and on such basic issues as conception of the self, morality, and conflict resolution" (Warren 1987, 10). In effect, "maternal feminists politicize these social psychological differences;" they see the woman citizen as "a loving being devoted to the protection of vulnerable human life and to making the virtues of mothering the template for a new, more humane public world" (Tilly and Gurin 1990, 22).

As Orleck notes, "for many women in cultures around the world, motherhood is a powerful political identity around which they have galvanized broad-based and influential grassroots movements for social change" (1997, 7). When race and ethnicity are introduced into the model presented by maternal feminists, it can provide a useful lens for examining the political activism of women of color. For example, as Gilkes notes, women of color

> highlight the importance of women and their work for the creation of a just and more equitable society. Women [of color] bring three perspectives to community work that make them particularly rebellious. First, their experience in the society... Second, they see men's suffering and feel its effects in their own lives. ...The third and perhaps most important source of discontent is the effect of racial oppression in the lives of their children. ...It is in their roles as the principal caretakers of children that racial-ethnic women pose the largest political threat to the dominant society (1994, 242. Emphasis added).

In general, women are socialized into the role of caregiver responsible for the wellbeing and maintenance of the home and family, and by extension, the community. In fulfilling their role as nurturers, women find themselves practicing a politics of care, which in turn may lead to their own political awakening and to their exercising leadership within the policy arena. Indeed, this is what Orleck (1997) found in her recent examination of the "politics of motherhood." Her comments about the activists interviewed echo those of Gilkes, above.

> Some of the women...were formerly apolitical housewives, first moved to take political action by concern for their own children. For them, the radicalizing process involved a series of leaps from caring about their

> own children, to caring about other mothers and children, to engaging with pressing social and political issues of national and even international import. Becoming politically active changed these women's understanding of themselves as mothers and as individual adult women, affirming in them a sense of collective power, entitlement and moral authority (Orleck 1997, 4).

The transformation experienced by the mothers in Orleck's book sounds very much like the three step process of consciousness raising proposed by Tolleson Rinehart. First, the individual begins to identify with women, acknowledging common interests. Second, she notices disparities in how women are treated and feels this is unjust. Third, she recognizes that "the problems women face demand collective, political solutions and cannot be solved through individual efforts" (S.J. Carroll 1989b, 328. Also see Fowlkes 1992). The ensuing gender consciousness sparked in women undergoing this process is explained by Tolleson Rinehart as

> one's recognition that one's relationship to the political world is at least partly but nonetheless particularly shaped by being female or male. This recognition is followed by identification with others in the "group" of one's sex, positive affect toward the group, and a feeling of interdependence with the group's fortunes (1992, 32).

Gender consciousness is substantially different from Kaplan's notion of "female consciousness," in which women,

> emphasizing roles they accept as wives and mothers, also demand the freedom to act as they think their obligations entail. ...Far from being a biological trait, female consciousness develops from cultural experiences of helping families and communities survive (1997, 6–7).

The critical distinction between "female consciousness" and "gender consciousness" is apparent in Molyneaux's (1985) theoretical framework for explaining women's politicization. She argues that women are pursuing their "practical gender interests" when they respond collectively to protect the well-being of their families, racial/ethnic communities, or class. Under these circumstances, women are driven by their female consciousness to act, within their role as wife and mother, in order to maintain the status quo. Their efforts are seen as more "nurturant than political," however, because their activism conforms within their expected roles and they generally seek to fulfill basic human needs (Dabrowski 1983). Following resolution of the problem or dissipation of the threat, the women return to their former way of life. In this instance, their experience of participation in the political arena has failed to fully politicize them.

In contrast, personal and political transformation ensue when women are mobilized to pursue their "strategic gender interests." In this case, feminist- or gender consciousness develops as an outgrowth of women's collective organizing around practical gender interests such as affordable housing, comparable pay, quality daycare, or clean drinking water (Kaplan 1997; Basu 1995; Peterson and Runyan 1993; Cable 1992; Molyneaux 1985). Once they are transformed by their experience of activism, the women cannot go back to the way they were. They have been changed, politicized, and will never see their world in the same way again.[9]

FEMINISM AND THE NEW MEXICO LEADERS

> I'm a rebel, not a feminist. I ask "Why?" when someone says women can't. I was the oldest in the family. I helped my father with the farming. I rode horses. Women in the community said "You're not supposed to do that" and I said "Why not?" (Toni, activist).

> I come from five generations of feminists. ...On the Comanche side of our family, we have to fight the tendency to think men are inferior to women [laughs] (Cora, public official).

The belief that all women share a common experience of oppression and subordination—based on the sexual division of labor and the organization of sexuality and procreation—underlies much of radical feminist theory (Young 1988). In particular, feminist thought generally shares

> a belief that women's opportunities and the quality of their lives are limited in part simply because they are women,...a desire to release the constraints imposed by patriarchy and sexism, and...a belief not simply in individual action, but also in group action to improve the conditions of women (Sapiro 1981, 710).

However, indigenous and Hispanic women's activism is often dedicated to advancing the quality of life for all members of their communities rather than advancing the position of women, in particular. One scholar posits that "some women's struggles form a subset of struggles for civil rights and human rights. What initially motivates many women to organize is not necessarily a belief in the distinctive nature of their problems but rather a sense of shared oppression with other groups that have been denied their rights" (Basu 1995, 10).

For instance, identity for some American Indian women is deeply rooted in their relationships to their extended family, their culture and tribal ties, and their spirituality (Hoikkala 1995). As women within their native urban or tribal community, they are

> fulfilling harmonious roles in the biological, spiritual, and social worlds... Biologically, they [value] being mothers and raising healthy families; spiritually, they [are] considered extensions of the Spirit Mother and keys to the continuation of their people; and socially, they [serve] as transmitters of cultural knowledge and caretakers of their children and relatives (LaFromboise et al. 1990, 457).

Indigenous women's support for and appreciation of motherhood and service to others does not preclude their support for women's economic parity or legal rights. However, the devaluation of woman's nurturing role by feminism, whether perceived or actual, has served to undermine the potential support of culturally traditional women. Perceptions of feminism's hostility toward the family and religious institutions have resulted in the alienation of many women of color from the women's movement. And, even those women of color attracted to feminism believe that the women's movement has been unresponsive to their specific concerns. As a result, many women of color reject the term "feminist" despite their agreement with the general principles of feminism. They face a dilemma: they are cognizant of the benefits of the women's movement, yet they recognize the limited potential of the movement for their particular cause.

Their predicament became apparent during the nationalist movements of the 1960s and seventies as Hispana and native feminists struggled to balance the often incompatible demands and ideologies of the women's and Chicano or American Indian movements.[10] As one native leader said of the women's movement,

> There were crazy white women wanting to do what men do. I didn't get involved in that. Being a macho woman doesn't fit the way I do things. I didn't have anything to prove (Valerie, public official).

Feminism seemed irrelevant to Valerie and other Indian women like her who felt that gender equity was less important an issue to her people than issues such as tribal sovereignty, public health, environmental quality, and economic development. She went on to explain,

> Our clans are matrilineal. You belong to your mother's clan. Everything is along the mother's line. And the most sacred thing to us is the corn. And corn represents the female, corn is the mother. So everything is female oriented pretty much... The reason why you hear a lot of emphasis on the female is because we're just trying to revive it and remind everybody that she was always there and after having been overshadowed by the male for so long we have to remind ourselves that one cannot be without the other (Valerie, public official).

In contrast to the mainstream Anglo-American women's movement native women organized around issues important to Indian communities. As a result, the North American Indian Women's Association, as well as other native women's organizations, formed to address concerns such as education, economics, tribal sovereignty, treaty rights, and health care. In 1977, at the National Women's Conference, Indian women developed the following manifesto.

> American-Indian and Alaska Native Women: have a relationship to Earth Mother and the Great Spirit as well as a heritage based on the sovereignty of Indian peoples. The federal government should guarantee tribal rights, tribal sovereignty, honor existing treaties and congressional acts, protect hunting, fishing, and whaling rights, protect trust status, and permanently remove the threat of termination. Congress should extend the Indian Education Act of 1972, maintain base funding of education instead of replacing it with supplemental funding, provide adequate care through the Indian Health Service, forbid the systematic removal of children from their families and communities, and assure full participation in all federally funded programs (quoted by Medicine 1978, 93).

Furthermore, for women of color, feminism appeared to ignore the historical rifts that separated women of differing classes, racial/ethnic groups, and cultures by simply proclaiming women's "sisterhood" and assuming their solidarity (Cotera 1980. Also see Basu 1995; Cruz Takash 1993; Pesquera and Segura 1993; Bonilla-Santiago 1991; Johnson-Odim 1991; hooks 1981). Women of color were keenly aware that, like them, their men also faced racial subordination and class oppression. Indeed, these women sometimes found themselves facing classism and racism in one camp, and sexism in the other (Garcia 1989). Consequently, they believed that a successful political movement must address the issues of race and class, and gain the support of the men in their community. This situation led to women of color incorporating their own perspective and aspirations for change into a feminism that better reflects their particular social, cultural, and economic condition. As one Latina scholar explains:

> Chicanas believed that feminism involved more than an analysis of gender because, as women of color, they were affected by both race and class in their everyday lives. Thus, Chicana feminism, as a social movement to improve the position of Chicanas in American society, represented a struggle that was both nationalist and feminist (Garcia 1989, 220).

For feminism to be a major resource for Hispanas and native women, its objectives and ideology must correspond much more closely with theirs. More to the point, women of color such as Hispanas and native women must first articulate their own goals and objectives in relation to their cultures, and they have to relate them to their distinct ideology (Junn 1997; Pesquera and Segura 1993; Trask 1993; Jaimes 1992; Aragón de Valdez 1980). Second, the women's movement must incorporate fully the needs of working-class women and women of color into their agenda (Basu 1995; Cable 1992; Bonilla-Santiago 1991; Johnson-Odim 1991; Cotera 1980). bell hooks (1981) envisions a new feminism, broader in scope, that speaks to the experience of individuals across sex, class, and racial lines.

> [F]eminism is not simply a struggle to end male chauvinism or a movement to ensure that women will have equal rights with men; it is a commitment to eradicating the ideology that permeates Western culture on various levels—sex, race, and class, to name a few—and a commitment to reorganizing U.S. society so that the self-development of people can take precedence over imperialism, economic expansion, and material desires. ...I choose to re-appropriate the term "feminism," to focus on the fact that to be "feminist" in any authentic sense of the term is to want for all people, female and male, liberation from sexist role patterns, domination, and oppression (hooks 1981, 194–195).

Since both Hispanas and American Indian women have historically experienced colonization, racism, economic marginalization and political repression in addition to sexism, one might expect that the New Mexico leaders would uniformly reject mainstream middle-class, white women's feminism. Nevertheless, this was not the case. Instead, many of the leaders chose to adapt or reformulate elements of feminist ideologies to better reflect their own experiences of what it means to be an indigenous woman or an Hispana living in the dominant Anglo American culture.

> Yes, [I'm a feminist] but not in the mainstream feminist movement. The white feminist movement doesn't share my goals. Some intersect, like the environment, but social justice issues are [the main concerns] of women of color. Gender is such an important undercurrent. ... It's very difficult when you do go to school because you're opening your mind and you're seeing things and you're being able to articulate the power structures. As you become empowered (and I think education is a real way to empower yourself), you're being able to debate and to contextualize where you're at in society, within your own group, and out of the group, and you're challenging and finding the women's movement in its own way (Margarita, activist).

In the course of their interview, each leader was asked whether she considered herself a feminist. Whenever a participant sought a definition for the word "feminist," which happened roughly a quarter of the time, I asked her to explain to me what she thought the term "feminist" meant, and to answer the question based on her understanding of the concept. For example, one leader said, "I'm not a feminist. It means your agenda should just surround family stuff, mother stuff, domestic violence stuff, money for human service programs. ...I deal with more than that" (Leonora, public official). This open-ended approach enabled leaders to articulate what they understood a feminist or feminism to be while providing a context for their answers.

The findings reveal the complexity of the leaders' conceptualizations of feminism and, occasionally, their ambivalence towards a highly charged term that clearly means different things to different people. For instance, when asked whether she was a feminist, one woman haltingly answered,

> In the traditional context, probably. Yes. It depends. It's a hard concept. In my interpretation, feminism means all people should have an equal voice, equal power, equal pay. Some people think the feminist movement is anti-male. But the world needs to work with men too. We all have to get along (Lucia, activist).

Overall, the New Mexico leaders fell into four general groups, those who clearly stated that they were not feminists, those who were ambivalent about feminism, those who readily self-identified as feminists, and those who self-identified as "Third World" or "indigenous" feminists.

Table 17. Leaders' Identification with Feminism

	Activists		Officials			
	Indig.	Hisp.	Indig.	Hisp.	Total	%
Not a "Feminist" [1]	5	3	4	3	15	30
Ambivalent Feminist[2]	0	1	2	0	3	6
Feminist[3]	5	5	4	8	22	44
Third World Feminist[4]	3	4	3	0	10	20

[1] Rejected the term "Feminist" to describe herself.
[2] Did not self-identify as a "Feminist," yet supported and/or promoted women's rights and policies that advance the position of women in society.
[3] Self-identified as a Feminist.
[4] Advocated a feminism that reflects the varied cultures, beliefs, and life situations of Third World women, including Hispanas and indigenous women.

As indicated by Table 17, above, the majority of the leaders (32 or 64 percent) acknowledged that they were either feminists or third world (or indigenous) feminists. Only three leaders (6 percent) were ambivalent while 15 (30 percent) rejected feminism and/or the term "feminist" to describe themselves. While a minority of the indigenous women (9 or 35 percent) identified themselves as feminists, a majority of the Hispanas (13 or 54 percent) did. And, although only 38 percent (10) of the grassroots activists described themselves as feminists, 50 percent (12) of the public officials did.[11]

The larger percentages of public officials and Hispanas embracing feminism are consistent with greater numbers of these leaders displaying gender consciousness (92 percent and 100 percent, respectively), using their activism to address women's concerns (50 percent and 58 percent, respectively), and becoming motivated to enter politics by their desire to improve the position of women in society (63 percent of both public Officials and Hispanas). (See Table 16). The 22 self-described feminists tied feminism to a belief in the strength, equality, and capability of women in both public and private life.

> If you mean equity for women, I am [a feminist]. I believe women can be as strong, as intelligent as men, and can carry out their duties as well as men. I believe women should be given the opportunity, just as men are, to compete on a fair playing field (Victoria, public official).

Being a feminist sometimes required courage, especially for women living in highly conservative Pueblos in which religion was closely tied to politics and women's participation was either limited or prohibited. As one leader explained,

> It's very difficult to be a feminist Pueblo woman. [Being a feminist means] taking an active role in positive change for women, not being afraid to stand up for what's right. Regarding tribal governments where women are not involved in decision-making or advice-giving: women have lots of important things to say. They want to be heard more. [Feminists] want women to have more input in decisions affecting women and children. We can influence informally through our husbands. We have to do it in a very discreet way. But [smiling] one councilmember told me I'd be the first woman governor of the Pueblo (Daphne, public official).

It was apparent that for some leaders, being a native woman and a feminist implied that they were atypical of, or somewhat different from, other Indian women. To illustrate, when asked what being a feminist meant to her, one native leader replied,

> I'm very outspoken as far as being an Indian woman. A lot of people say "She's a feminist." I think I am, given my job here. I make more

> money, I can support myself. My idea [of a feminist] is being someone who's independent, who stands up for a position, and I think I've done that within this organization [and] within the tribal groups that I work with where it's primarily male-dominated. I'm a department head, and for the most part, they've always been men. So, in my realm, compared with other Indian women, I'd say yes, [I'm a feminist] (Karen, activist).

All 22 feminists exhibited gender consciousness by identifying with other women as a group and supporting the advancement of women politically, socially, and personally. Not all of the gender conscious New Mexico leaders thought of themselves as feminists, however.[12]

While it was common for the New Mexico leaders to display ambivalent feelings towards feminism, 47 of the 50 asserted that they either were not feminists or that they were feminists of one form or another. Only three leaders remained ambivalent towards feminism; one was an Hispanic activist (comprising 4 percent of the Hispanas and activists) and two were American Indian public officials (constituting 8 percent of the indigenous women and officials). Even when a leader rejected feminism personally, her ambivalence regarding its value as a movement was evident as Marta's comments reveal, below.

> I don't consider myself part of the feminist movement. Not because I'm against it, but because it's just not my issues. I sit on the Ms. Foundation board and I value the work they do. I think it's very important for middle class women to not take things lying down. And it can only make things better for all women. But those aren't my issues. My issues are the families. Our families...are under siege in our communities, whether they're urban or rural. The genocide that's happening with our young men, the pressure to assimilate Mexican Americans today, the pressure to lose land, those are my issues. And to the degree that anybody fighting those issues feels women are inferior, I'm going to fight them on it. And [if they] cannot accept leadership from women, I will not work with them. I identify with Anita Hill, even though that never happened to me. I believe her. So, I don't know what I am, but I can tell you what I'd do in certain situations (Marta, activist).

In addition to her involvement with the Ms. Foundation, Marta's allusions to Anita Hill, women's leadership, and her "fighting" anyone believing women to be inferior illustrate her support for feminist principles. At the same time, Marta saw feminism as a white middle-class women's movement that does not address the issues of greatest concern to her: the preservation of Mexican American people's culture and traditional lifeways. Marta's final statement, "I don't know what I am, but I can tell you what I'd do in certain situations," indicates this leader's awareness of the incongruity between her support of feminist goals and her reluctance to identify as a feminist. This contradiction emerged frequently in the New Mexico leaders' narratives, often as ambivalence towards feminism or as a tension between a woman's support for feminism and her loyalty to her racial/ethnic group. Scholars have noted that, "Resistance to feminism may reflect a fear that it demands a total transformation of the social order" (Basu 1995, 7) and the subsequent undermining of one's culture and traditions (Aragón de Valdez 1980).

Some of the leaders who rejected feminism claimed that they themselves had never experienced any form of sexual discrimination and that they had simply worked hard to achieve their leadership position. Yet the ambivalence felt by such women towards feminism is revealed by the conflicting emotions expressed in one former state legislator's narrative.

> You have to get out and work hard and accomplish things and forget that you're a female. You have to go in there and think like [men] do and work side by side and accomplish things. Quit crying that you don't get this because you're a female. You just have to go out and prove yourself. You have to work harder. When you first go in as a woman, especially way back in '75, there weren't too many women in politics. Everybody looks at you as a token. So you have to go out and prove yourself. ...I always helped the ladies when they had bills on women and stuff... [The lack of] equal pay for women bothered me a lot. You know women who do the same job as a man but always get paid a lot less. That was one of the harder things we had to accomplish (Alicia, public official).

Alicia's comments exposed her feelings of anger, resentment, bitterness, and pride as she acknowledged the difficulty women have faced breaking into traditionally male bastions of power, such as the state legislature, where she served for years. Believing that people are responsible for their own destiny, Alicia thought women should pull themselves up by their bootstraps. "Quit crying," she admonished, "you just have to go out and prove yourself." She recognized, however, that as one of a handful of women in the New Mexico legislature during the 1970s, she was seen "as a token." Although this public official rejected the label "feminist," Alicia cooperated with other women legislators on issues affecting women such as pay equity. In fact, while she initially distanced herself from the other women legislators—they were "the ladies"—she later admitted that getting pay equity legislation passed was "one of the harder things *we* had to accomplish."

Alicia's unwillingness to identify herself as a feminist, or even as one of the "lady" legislators while she actively supported women's concerns, was not unusual for someone in her position. Studies of women in public office reveal a general pattern of advocacy for a feminist agenda: one that incorporates policies or programs that promote equality or improve the status of women (Thomas 1994; Cantor and Bernay 1992; Le Veness and Sweeney 1987). However, these leaders are sometimes reluctant to openly identify themselves or their policy goals as feminist. Susan J. Carroll (1984) coined the term "closet feminist" to describe the ambivalence of such politicians towards feminism. In a 1992 study, Deutchman found that women legislators often believed that the label "feminist" was a liability to their political career. However, their failure to openly embrace feminism did "not indicate these legislators pursue anti-feminist or non-feminist policies; rather it is indicative of how complex women legislators' attitudes are toward their own minority status" (Deutchman 1992, 431. See also Prindeville and Braley Gomez 1999; Thomas 1997; Saint-Germain 1989; Antolini 1984; Boneparth 1984; Flammang 1984a). Further, Deutchman related the situation of the women legislators to that of women breaking into any male-dominated profession. More recent research indicates, however, that as the numbers of women in public office continue to increase, fewer women feel the need to remain "closeted" with regard to their feminist policy agendas (S.J. Carroll 1994).

While Alicia did not identify as a feminist or acknowledge experiencing personally any sexual discrimination, like other New Mexico leaders, she nevertheless showed strong gender identification. On the other hand, a few women admitted facing hurdles and prejudice simply because they were females active in politics.

> The last two years of [legislative] sessions, I was fired from my two jobs because I had to be at the legislature for 60 days. The next time it was 30 days for a special session. They didn't fire men working for the tribe who had the same involvement (Leonora, public official).

Even women aware that the injustices they experienced were purely due to their sex perceived that feminists were weak, whiny, anti-family, and/or man hating—an image that they did not want to be associated with and which hints at an underlying homophobia. These leaders felt that strong women could (and should) overcome whatever adversities they faced. As one woman noted, "I don't think a woman should be out with a pick and shovel. But on the other hand, if that's the only job, she better go out and take it" (Monica, activist). An intricate combination of feelings, ideology, and experience contributed to the seemingly contradictory statements made by women who rejected feminism, yet demonstrated a strong gender consciousness. Their conflicted feelings were, in part, a product of their adopting one set of beliefs and expectations (i.e. you can accomplish anything you wish to through hard work) while experiencing tremendous barriers to success such as injustice, racial prejudice, and/or sexual discrimination.

The discord between ideology (what should be) and experience (what is) encountered by many women is illustrated in a particularly moving narrative. Vivian, a professional with a pleasant, self-assured demeanor began speaking in a strong, confident voice and ended, hands trembling, with eyes full of angry tears.

> I'm a female. I don't believe I need to tear a man down to elevate myself. I was in charge of meetings, always in a male arena. I can hold my own with the best of them. [Being female] has never hindered me in my life. I was one of the highest paid women in government. The highest compliment I've ever been paid was when a supervisor said to me "You don't think like a woman, you think like a man." ...I know there's biases. I know that there's not equality in pay—all those things. The stats are out there. But it's never been anything that has held me back. I mean my career started off at the lowest level. I was a clerk, I worked my way up the organization to be the Director in charge of the _____ program. I firmly believe that in America, you go as far as you want to go. A lot of that comes from within you. Sure you fall flat on your face, and sure there are people that try to keep you down. I could write a book about what went on in my tenure in government and what I went through, and the hardships because I was a pioneer at a time when women were not in the finance field. I could tell you horror stories of some of the things that I was put through. And the sad commentary was that I was not just put through it by the men. The women didn't help me any either. They were very jealous because they were either in secretarial positions or in clerk/typist positions and I wasn't "one of them." I wasn't one of the guys either so I fell nowhere. So I went through quite a bit in my 25 years. If I have a grandchild that's female, she could look back and say: "My Grandma did this," and get encouragement. I didn't have that kind of encouragement. Women need to build networks to support, help, and nurture each other (Vivian, public official).

As she spoke, Vivian expressed great passion, sadness, anger, resentment, pride, and hope for the future. Although she started out claiming that her sex "has never hindered me in my life," Vivian soon revealed that she could "tell you horror stories of some of the things that I was put through" due to her being a "pioneer at a time when women were not in the [public] finance field." In answering the question "Are you a feminist?," Vivian illustrated how her socialization, political ideology, values, dreams, and life experiences were woven together into a rich, complex tapestry of contradictions.

For leaders such as Alicia and Vivian, feminism and the women's movement seemed unnecessary. According to their worldview, strong, responsible individuals form the basis of a healthy, mutually dependent, and caring community. They believe in the individual's talents and ability to overcome whatever obstacles are placed before her. These leaders may acknowledge that institutional, structural, and attitudinal barriers to women's advancement exist, but they refuse to see these obstacles as insurmountable. Rather than change the rules of the game to benefit women, however, they suggest that women learn to play by the (men's) rules, believing that women are capable of winning. These seemingly paradoxical perceptions of their experience in politics have been similarly observed in other female office-holders.

In a study of state legislators, Thomas (1997) examined how women coped with the pressures of being a (sometimes unwelcome) minority within the largely male arena of legislative politics. She found that about 25 percent of the leaders she interviewed

> dealt with the consequences of their dual status by dismissing the possibility that being a woman made any difference whatsoever to what they did and asserting that no one saw them any differently from men. ...[H]owever, almost every one of these women...subsequent[ly]... emphasized some way in which women were treated differently from men (1997, 33).

These so-called "Avoiders" tended to avoid issues primarily of concern to women and, in particular, feminist issues. While New Mexico leaders such as Alicia and Leonora tried to distance themselves from what might be deemed "women's issues," they encouraged the increased participation of women in politics and, when necessary, actively supported legislation especially of concern to women. Apparently, the desire of some women to be seen simply as leaders, rather than as *women* leaders, is strong among women in electoral politics. This is unlikely to occur, however, until women are accepted fully into all facets of politics and society and are no longer seen as intruders or "outsiders."

Despite her "think[ing] like a man," Vivian was neither fully accepted by her male colleagues nor by the women in "secretarial...or in clerk/typist positions" within her public organization. And, while Vivian proclaimed that being a woman has "never been anything that has held me back," she still urged women to "build networks to support, help, and nurture each other." Like Vivian, other leaders were similarly disappointed by the lack of support or the outright hostility they received from other women.

> I try to make an impact on women. We need more women in government at all levels, in the decision-making process. [I'm frustrated] that women's groups won't even come to me. Even women don't support me. Women are our greatest enemies. We don't network. We should take a lesson from men. Plan. Discuss. Don't criticize. We need to build agendas (Leonora, public official).

While Vivian and Leonora developed successful political careers on their own, without the help or direction of other women and despite much discouragement, they extolled the importance and value of having the support of other women. More fortunate leaders, such as Jacinta (a self-identified feminist), benefited from the mentoring and assistance of other women and praised those who had been instrumental in helping them achieve their political goals.

> We need to support one another. If it had not been for my cousin and two other women, I would not have gotten my first cabinet position. I've had a lot of women support me over the years. And I've tried to do the same with others. I believe that women need to continue to fight for what they don't get because they do good work (Jacinta, public official).

While some of the New Mexico leaders repudiated feminism, believing it to be incompatible with their roles as wives and mothers, others thought that it excludes the voices and concerns of women of color. All in all, nine indigenous women (35 percent), six Hispanas (25 percent), eight grassroots activists (31 percent), and seven public officials (29 percent), declared that they were not feminists. Despite their lack of identification with feminism, however, several (7 or 34 percent) of the 18 leaders (ambivalent- and non-feminists combined) promoted actively women's issues, sought to broaden social, political, and economic opportunities for women, and/or mentored other women specifically to aid their entry into leadership positions.[13]

The final group of ten women also rejected the white middle-class feminism of the women's movement but alternately proposed a brand of feminism incorporating their own culture and beliefs. These "third world feminists" and "indigenous feminists" included six indigenous leaders (23 percent), four Hispanas (17 percent), seven grassroots activists (27 percent), and just three public officials (13 percent).[14] Their perspectives of feminism generally echoed Leslie's words, below.

> When I hear the word "feminist" it usually has a Western definition. I'm not a feminist from [that] standpoint. ...We have a concern with our communities and families first. ...I know there are a lot of issues that need to be addressed but we look at them from a community standpoint, as a whole (Leslie, activist).

As we found in the discussions of racial identity, for some women, the good of the collective overshadowed the needs of the individual. Fighting racism and struggling for the greater social and economic good of the community were greater priorities than pursuing concerns specific to women. Juana, for instance, rejected feminism because "white, middle-class feminism is racist." Feminism had not represented her particular interests, as a woman of color, or her community's concerns. Instead, Juana described herself as a "third world feminist" because she believed that cooperation between men and women is essential and that "there is room for male and female leadership, partnerships." Furthermore, Juana felt an affinity with men of color explaining that she had "found that most third world men understand [sexism because] they have felt racism" (Juana, activist).

Juana, and other women like her, identified themselves as "indigenous feminists" or "third world feminists" to emphasize their distinct political and social position as women and as colonized racial/ethnic minorities. In a similar vein, Salina distinguished herself from middle-class white feminists by specifying that she was "a Chicana Feminist, real strong in who I am. Strong culturally, with feminist ideals" (Salina, activist). She went on to say,

> I don't consider myself allied with the white feminist movement. There are differences in the way we define feminism. [I'm a feminist] in the sense that I'm not a traditional Chicana—in the way we structure our household, in the way our relationship works out, in terms of my compañero [male companion]. Even in the organizing I do, I try to push a feminist agenda. In the organization, I look at ways we can involve more people: having meetings in the evening, providing childcare, that kind of stuff. Bringing up feminist dialogue in our organizational meetings because it's a good way to educate ourselves about our own internalized oppression. There are real obvious reasons why we need to be feminist. We're still earning 60 cents to the dollar. The whole situation. The world would be much better if more women were in leadership positions. I have a lot of strong women friends. Women have a more nurturing orientation, we're much more helping... I think even in leadership development and empowerment, women naturally know how to do that more. We know how to provide structures for dialogue amongst us (Salina, activist).

While leaders like Juana and Salina did not associate themselves with the white feminist movement, neither did they reject feminism completely. Rather, they saw value in its principles and opportunities to "take the meat from the bones," applying what was meaningful to them (and their communities) and discarding the rest.

The differences between the feminist ideologies of self-identified feminists and those who distinguished themselves as "Chicana-," "third world," or "indigenous" feminists has been similarly observed in a study of 153 Mexican American women. The researchers developed a typology of Chicana feminisms, which included what they termed "Chicana liberal feminism" and "Chicana insurgent feminism" (Pesquera and Segura 1993).

These forms of feminism roughly equate with liberal feminism and socialist feminism, respectively.[15] With regard to the New Mexico leaders of my study, the self-identifying feminists parallel the Chicana liberal feminists, and the third world feminists are similar ideologically to the Chicana insurgent feminists. To illustrate, Chicana liberal feminism is described as

> center[ing] on women's desires to enhance the well-being of the... community, with a special emphasis on improving the status of women. This perspective advocates change within the liberal tradition emphasizing access to social institutions, employment and equal treatment in all areas of life (Pesquera and Segura 1993, 106).

On the other hand, Chicana insurgent feminism

> emanates from a political tradition that contests the social relations of production and reproduction. Women who articulate these sentiments call for...self-determination which encompasses a struggle against both personal and institutional manifestations of racial discrimination, patriarchy, and class exploitation (Pesquera and Segura 1993, 106).

Like Pesquera and Segura's insurgent feminists, the New Mexico third world feminists transformed traditional (liberal) feminism to more accurately reflect the realities of working-class and third world women's lives. This pragmatic approach to reappropriating feminist ideology has been noted elsewhere in the literature. For example, in her study of women's movements worldwide, Basu similarly found that "many women who believe that feminism is bourgeois or Western go on to identify indigenous alternatives to Western-style feminism within their own cultural and political contexts" (1995, 19–20). Pardo calls this process "decolonizing feminism," by which she means "rethinking conceptual frameworks and political agendas and redefining and widening feminist terrain to include other bases of oppression such as race, class, and imperialism" (1995, 358). In this way, women of color reappropriate the women's movement by tailoring feminism to fit their own and their community's particular needs and situation.

Tolleson Rinehart (1992) reminds us, however, that while women of color have been wary of feminist cultural imperialism, racial differences may be reconciled by the commonalities of women's gendered experience. As S. J. Carroll notes, commonality among women

> may stem from the fact that women as socially constructed, gendered beings have certain shared interests resulting from their position in the sexual division of labor and the oppression they face even though the strength and nature of the sexual division of labor as well as the man-

> ifestations and circumstances of their oppression may vary considerably depending on race, class, sexual orientation, and other cultural differences (1992, 26).

If gender consciousness is accepted as a construct large enough to incorporate all permutations of feminism and a wide spectrum of women's political behaviors, it is reasonable to suggest that native and Hispanic women can reject the label of "feminist" while still being quite gender conscious. Or, from the perspective of some women of color, feminism represents an elitist, white women's cause, while gender consciousness does not dilute the racial dimension of their activism. Gender consciousness respects the fact that "women have bonds other than their bonds to women: to their race, to the cultures of their home places, to their sociodemographic environment... Thus we see that gender consciousness is a matter not only of gender identification, but of role ideology as well" (Tolleson Rinehart 1992, 70). Whether or not they embraced feminism, all 50 of the New Mexico leaders advocated legal, social, and economic equity for women. Perhaps what is needed, is a greater acknowledgment of the many feminisms that exist. As Pardo suggests,

> a multidimensional definition of feminist groups may provide for a more dynamic and contextual concept that involves class, ethnic, and gender struggles. If we open the borders of feminist frameworks and theories, we may broaden, strengthen, and enrich feminist political agendas and equate women's rights with other human rights (1995, 370. See also Pesquera and Segura 1993).

Similarly, conceptualizations of "environmentalism" may be broadened to include the values and political objectives of women of color who are concerned with the health and well-being of their people and with the quality of life of their communities. We now turn to a discussion of environmentalism as it was understood and practiced by the New Mexico leaders.

ENVIRONMENTALISM AND THE NEW MEXICO LEADERS

> I believe that as a human being it's my responsibility to be a good steward of the earth and help take care of her. And that when we take her resources we leave her the way we found her. And that we don't pollute our water, our air, our earth for those who come after (Sonia, public official).

As we can see from Table 18, only a minority (21 or 42 percent) of the 50 New Mexico leaders called themselves environmentalists. Just over half of the Hispanic leaders (13 or 54 percent), one-third of the native leaders

(8 or 31 percent), 50 percent of the public officials (12) and 35 percent of the grassroots activists (9) identified themselves using this label.

Table 18. Leaders' Identification with Environmentalism

	Activists		Officials			
	Indig.	Hisp.	Indig.	Hisp.	Total	%
Environmentalist	4	5	4	8	21	42
Non-Environmentalist	1	0	0	1	2	4
Third World Environt.[1]	8	8	9	2	27	54

[1] Third World Environmentalist: Advocates an environmentalism that incorporates participatory decision-making and locally determined priorities that reflect the concerns and values of indigenous and third world peoples.

The "environmentalists" demonstrated mainstream environmentalist values such as preservation of wilderness areas and maintaining biodiversity, and saw the appropriate role of humans as being stewards of the earth. These leaders sometimes worked in cooperation with and/or supported the work of national environmental organizations, approaching policy change from within the existing political framework rather than challenging established social, political, or economic structures.[16] This meant that these self-identified environmentalists believed in the capacity of (and employed) the existing political and economic systems to promote environmental quality within their communities. As one leader stated, "I believe we need to take care of the environment for ourselves and our children. We've seen the benefits of regulations and major legislation such as the Clean Air and Clean Water Acts" (Gina, public official).

As in the case of feminism, however, several of the New Mexico leaders objected to the white middle-class bias they believed was inherent in mainstream environmentalism.

> I never said I was an environmentalist, but I believe in the environment. ...I believe in preserving, in respecting the use of resources. Respect for the environment is essential. If you take care of the earth and you take care of the heavens—important things in life—that will always be there to sustain your life (Erica, activist).

Despite some of these leaders' involvement in or association with environmental groups, for the most part, people of color do not join such organizations for recreational or aesthetic purposes (Bullard 1993; Taylor 1993; Freudenberg and Steinsapir 1991; Alston 1990). This is not to suggest that they are not interested in the environment; on the contrary, they may be very concerned with maintaining the health and safety of their communities. For example, one native leader said,

> I've worked real hard on environmental issues. I've brought the state and tribes together on environmental issues; for example, on solid waste. There are no borders where waste is concerned, we must work together (Lorna, activist).

Ironically, because racial/ethnic minorities are more likely to experience undesirable or even hazardous living conditions (due to pollutants emanating from nearby industries or waste dumps, for example), they are unlikely to associate their struggle with that of environmentalists.

Consequently, over half of the New Mexico leaders (29 or 58 percent) did not consider themselves "environmentalists" even though they espoused values that included respect for nature and preservation of the earth's resources. Instead, the majority (27 or 54 percent) expressed the values and goals of the environmental justice movement describing themselves as "new," "third world," or "indigenous" environmentalists.

> An indigenous environmentalist is one who believes in the spiritual value of nature. One who sees nature not only in a patch of forest, but also in the middle of downtown. An indigenous environmentalist does not see anything "wild" about nature (Linda, activist).

Specifically, 65 percent of the indigenous women (17), 42 percent of the Hispanas (10), 62 percent of the grassroots activists (16), and 46 percent of the public officials (11) identified themselves as, what I will call, a third world environmentalist. Only two women (4 percent) failed to identify as either environmentalists or third world environmentalists. One of these "non-environmentalists" was an indigenous activist concerned with the effects of water pollution on the health of her tribe's cattle, and the other was an Hispanic official who believed that humans have a responsibility to "respect what the Lord has given us."

As participants or supporters of the environmental justice movement, the third world environmentalists differentiated their perspective from that of the mainstream environmental movement claiming that the latter is out of touch with the economic and social realities of the poor, the working-class, and of people of color. Di Chiro, who also interviewed women of color in the environmental movement, similarly found that they were reluctant to call themselves environmentalists "due to the dominance of the mainly white, middle-class, and uncritically 'preservationist' political culture from which much of the mainstream environmental discourse has developed" (1992, 94). The principal differences between the New Mexico leaders interviewed and those in the mainstream environmental movement were their approaches, their constituencies, and philosophies. Proponents of the environmental justice movement were concerned with the economic and political empowerment of third world communities. "Environment" was defined broadly to include "where we live, work, and play."

Indeed, the problems that third world environmentalists face are far more immediate and personally threatening than the traditional environmental concern for preservation of wilderness areas and nonhuman species, which, they regard, of less consequence than the survival of their people and their communities (Austin and Schill 1994; Marquez 1994). The New Mexico leaders who identified with the environmental justice movement frequently advocated incorporating the cultural beliefs of third world peoples relative to sustainable use (conservation) into contemporary environmentalism. Marta's comments, in particular, illustrate this view.

> When you have settlers that have come onto the land and have survived more than a century, if they came to settle as farmers and ranchers, they will have developed a relationship to the resources in that ecology that they have had to respect [in order to] survive in very harsh climates, which we have. The science that gets learned over several generations, in terms of how people relate to the land and its resources to keep it healthy, is a science that is today being thrown out the window by environmentalists and the government. So keeping the culture strong is an environmental mission (Marta, activist).

Belief in the value of indigenous people's knowledge and experience of the earth is central to the environmental justice movement. Winona LaDuke, an indigenous environmentalist, ties environmental protection to cultural preservation and economic autonomy for her tribe (Orleck 1997). The New Mexico leaders often characterized the difference between the environmental justice movement and environmentalists as an issue of utility over ideology. As one Hispana explained of her political involvement,

> I got rid of some of my friends. A lot of people think I'm crazy. I've dropped those who don't have the same thirst for justice, for the people. I believe in ecology, in the restoration of the human spirit, of the earth. I have friends who are environmentalists and we've agreed to disagree but sometimes it's very uncomfortable. It's a touchy subject but people need to work to eat. We have to have jobs, responsible industries. You can't rely on cottage industry (Dalia, activist).

While the Hispanas and public officials were somewhat evenly split between environmentalists and third world environmentalists, approximately two-thirds of the indigenous women and grassroots activists identified as third world environmentalists. The perspectives of the environmental justice movement shared by the third world environmentalists included: a strong belief in the rights of citizens to participate in making environmental decisions; a general distrust of government based upon direct experiences with public officials and agencies; a basic belief that human health—rather than aesthetics, wilderness preservation, or other issues—is their primary concern; a skepticism about science and industry; and a prevalent belief that economic growth is not necessarily good and does not benefit everyone equally. Several New Mexico leaders also shared the opinion that "our land is being ripped-off with the help of the top ten mainstream environmental organizations that have colluded with industry, the military, and government, and agribusiness" (Juana, activist). These leaders' involvement in environmental issues was often highly personal because threats to the environment were interpreted as threats to their families and communities.

For women of color, the political activities associated with environmental justice are predicated on the critical linkage between race and the environment rather than class and the environment. Leaders such as Juana saw toxic contamination of their communities as systematic genocide believing that communities of color are being targeted by government and industry for environmentally undesirable land uses that will result in the annihilation of their people. The term used within the movement to describe this practice is "environmental racism." A review of the literature indicates that environmental racism is a recurring theme in many of the narratives of American Indian women activists; as one scholar writes, "it is a genocidal analysis, rooted in the Native American cultural identification, the experience of colonialism, and the immanent endangerment of their culture" (Krauss 1993, 257. See also Gedicks 1993; Churchill and LaDuke 1992; Jaimes 1992; Gunn Allen 1986).

To better understand their conceptualizations of "environmentalism," I examined the basis of the New Mexico leaders' environmental beliefs. In general, these fell within one of four categories: preservation, sustainable use or conservation, environmental justice, and spiritual beliefs.[17] As the leaders' environmental ethics were often richly described and conceptually complex, their beliefs were seldom contained within just one category. Therefore, majorities expressed environmental beliefs that straddled, for example, preservation, sustainable use, and environmental justice. (See Table 19, below). Rather than signaling their lack of commitment to one point of view, the findings reveal the multi-faceted nature of the leaders' environmental ideologies, their appreciation for the complexities involved in seeking solutions to difficult environmental problems, and their understanding of the tremendous cultural variation of perspectives of the environment.

Of the 50 New Mexico leaders, 35 (70 percent) based their environmental beliefs, in part, on preservation. Like mainstream environmentalism, "preservation" stressed the value of maintaining wilderness areas and biodiversity—not only for the enjoyment and use of future generations—but because nature has value in and of itself.

Table 19. Basis of Leaders' Environmental Beliefs

	Activists		Officials			
	Indig.	Hisp.	Indig.	Hisp.	Total[1]	%
Preservation[2]	12	4	11	8	35	70
Sustainable Use[3]	11	12	11	8	42	84
Environmental Justice[4]	13	13	5	2	33	66
Spiritual Beliefs[5]	10	3	8	3	24	48

[1] Leaders indicated multiple bases of their environmental beliefs; these are not ranked in any way.
[2] Like mainstream environmentalism, preservation stresses the value of maintaining wilderness areas and biodiversity for their own sake, whether or not their existence directly benefits human beings.
[3] Sustainable use or conservation involves the development and "wise use" of natural resources and places human needs above those of non-human life.
[4] Environmental justice incorporates social, economic, and racial- and gender equity goals. "Environment" is defined as "where we live, work, and play."
[5] The natural environment and non-human life are sacred and valued for their own sake; all life is seen as an interconnected web.

The role of humans was seen as stewards of the earth. A large majority of the indigenous leaders (23 or 88 percent), in contrast to half of the Hispanas (12), 62 percent of the grassroots activists (16), and 79 percent of the public officials (19) supported preservation. While some of these leaders identified as mainstream environmentalists, others did not, even though they favored aspects of preservation. Some leaders disagreed, for example, with the approaches taken by national environmental organizations; others were concerned with the impact of specific policies such as the endangered species act on local economies. Nevertheless, these New Mexico leaders shared a belief in the innate value of nature.

The difference between preservation and conservation was not always clear cut, as is evident from the following quote.

> [An environmentalist is] someone who takes an active role in preserving many things including culture, nature, people. [Someone] who looks at the world like we all fit together for a reason. If we lose the Mexican wolf or a plant species then we'll lose indigenous people. There is an interconnectedness (Lucia, activist).

While this leader called herself an environmentalist (rather than a third world environmentalist), she sought to protect both endangered species and the lifeways of human communities indigenous to New Mexico. She did not place greater value on either humans or animals but rather saw the need for preserving the balance among all living things. Lucia, and other leaders like her, incorporated elements of both preservation and conservation into their environmental belief systems. Such leaders were able to bridge effectively the political and ideological differences between the mainstream environmental groups and local grassroots organizations that worked for environmental justice and advocated sustainable use.

Sustainable use largely placed human needs above those of non-human life and was the environmental approach preferred by 84 percent (42) of the New Mexico leaders. Whereas half of the Hispanas had supported preservation, 83 percent (20) based their environmental beliefs on sustainable use as did 85 percent of the native leaders (22), 88 percent of the activists (23), and 79 percent of the officials (19). "Sustainable use" referred to the conservation of land and natural resources, and implied a preference for environmentally sustainable forms of economic development (Paehlke and Rosenau 1993). Conservation policies were, in many cases, both economic and political as they sometimes pitted local interests against those of business and/or the federal government.

For example, local ranchers and sheep herders have engaged in numerous disputes with the U.S. Forest Service, the state Game and Fish Department, and environmental groups over livestock grazing (Peña and Gallegos 1993; Pulido 1993). Ranchers, who wield significant political clout within the state and are largely Anglos, generally pay below-market grazing fees when grazing their herds on public lands. On the other hand, *pastores* [sheep growers] within the state, most of whom have small operations and are Hispanic, have either lost their traditional communal grazing lands (Spanish land grants), or have sometimes been denied the same access to public lands for grazing their sheep that the cattle growers have enjoyed. The *pastores* see themselves as keeping an important cultural tradition alive while contributing to the economy of a significantly depressed region within the state. The cattle growers see the *pastores*' sheep as unwelcome competition for fodder. Environmental groups perceive both the ranchers and the sheep growers as contributing to the destruction of public lands. The political significance of this complex situation is illustrated, in particular, by Magdalena's comments.

> My...father raised sheep for a living so I think the environmentalists have gone too far as far as grazing... Some of the things they've done to the ranching families; [shakes her head] the people up north who are not allowed to graze their sheep and their cattle. You know, the Indians talk about the broken treaties; the United States also took the land from the Hispanics. Over 80 percent of the Hispanics that had land grants and grazing rights lost them. And it has been to our detriment

> since becoming second class citizens of the United States (Magdalena, public official).

The fact that environmental groups have allied with governmental agencies against local Hispanic communities has exacerbated the tension between these groups, as is clear from Marta's statement.

> [Mainstream environmentalists] are economically and biologically illiterate; they do not understand local conditions. They take off on their fantasies about a pristine type environment and what it should look like and try to apply that to every piece of land. ... "Trickle down" environmentalism is "If we litigate and if we legislate, it will trickle down and it will create a healthy environment." And just like trickle down economics, it does not get rid of poverty (Marta, activist).

The overwhelming support for conservation by the New Mexico leaders may be explained, in part, by its strong cultural component. Unlike Indian tribes in other states, the Pueblos were not displaced from their lands by the federal government and, therefore, they maintain an historical connection to the land. Furthermore, some Hispanic families have managed to maintain portions of their Spanish land grants despite significant pressure to sell. Proponents of sustainable use emphasized the historic ties to the land that both native people and Spanish descendants share. Historic practices such as ritual field burning to replenish the soil, the maintenance of *acequias* [irrigation canals], sheep herding and weaving, and the cultivation and use of indigenous medicinal plants by *curanderas/os* [healers] are rich cultural traditions unique to New Mexico. "Sustainable use" advocates maintenance of traditional agrarian and/or pastoral lifeways, which was seen as especially important to this group of leaders, for preserving New Mexico's indigenous and Spanish cultures. Such views tie in directly with the political goals of the environmental justice movement; to obtain social justice and economic parity for communities of color.

Of the women interviewed, 66 percent (33) based their environmental beliefs, in part, on principles fundamental to the environmental justice movement. Specifically, 69 percent of the indigenous leaders (18) and 63 percent of the Hispanas (15), all 26 of the grassroots activists, but only 29 percent of the public officials (7) displayed an environmental justice perspective. This meant that their environmental beliefs incorporated issues relative to social and economic justice, and included racial- and gender equity goals in addition to improving environmental quality. For example, environmental justice implies sharing both the costs and benefits of environmental policies so that no one racial/ethnic or socioeconomic group assumes disproportionate rewards or burdens (such as the cost of hosting a toxic dump). As one activist explained,

> The grassroots environmental justice movement's focus is on survival. Economic needs must be addressed. The environmental movement is more leisure and recreation-oriented, they deal with different issues. For example, we need trees for firewood, land for grazing. We see [the environment] as one whole thing. The environment reflects on our spiritual development—protecting the earth for our survival (Terri, activist).

Since all of the grassroots activists interviewed were engaged in politics around issues of environmental justice, it was no surprise that each of these women embraced such environmental beliefs. It was unexpected, however, that fewer than one-third of the public officials incorporated environmental justice ideals into their environmental belief systems; especially since 46 percent of the officials fit the category and/or referred to themselves as third world environmentalists. This apparent inconsistency may simply reflect their adoption of aspects of the environmental justice ideology, which they tempered with support for preservation and sustainable use, environmental approaches that were more representative, perhaps, of a broader, more environmentally moderate constituency. After all, these leaders chose to enter electoral politics, a more conservative arena than that of grassroots activism, where even environmental justice is perceived as radical for its calls for economic equity and its charges of environmental racism. In any case, as noted in Terri's comments above, the New Mexico leaders' spirituality and religious beliefs sometimes entered into their environmental ideology as well.

Nearly half of the New Mexico leaders (24 or 48 percent) included spirituality as a basis for their environmental beliefs. One-quarter of the Hispanas (6), half of the grassroots activists (13), and 46 percent of the officials (11) displayed an environmental ideology interwoven with their spiritual beliefs. However, as might be expected, a majority of the indigenous leaders (18 or 69 percent) exhibited spiritual-environmental beliefs. As Gunn Allen writes,

> If American society judiciously modeled the traditions of the various Native Nations, the place of women in society would become central, the distribution of goods and power would be egalitarian, the elderly would be respected, honored, and protected as a primary social and cultural resource,...the destruction of the biota, the life sphere, and the natural resources of the planet would be curtailed, and the spiritual nature of human and nonhuman life would become a primary organizing principle of human society (Gunn Allen 1986, 211).

"Spiritual beliefs," relative to the environment, were grounded on the notion that both the natural environment and all non-human life within it are sacred and, therefore, are valued for their own sake. Life was seen as an interconnected web in which maintaining a harmonious balance is essential to its integrity. One woman explained that,

> [Environment] has a different meaning for most Indian people. We can't separate it from our everyday being. You grow up learning and being a part of the environment; not a separate entity. I find it real sad that people separate the words "Indian" and "environmentalist." To me it's one and the same. Like being a civil rights activist and an Indian. They're synonymous[18] (Carmen, activist).

Carmen's statement implies that there is a political component to her environmental beliefs and to her identity as an American Indian woman. The interconnectedness among native spirituality, environment, and tribal sovereignty has been prominently illustrated by an ongoing struggle between development interests and native peoples in the state. In recent years, Indian tribes across New Mexico have allied with both grassroots environmental justice organizations and national environmental groups to litigate against the U.S. Forest Service, the Bureau of Land Management, and municipal authorities to protect religious sites such as Petroglyph Park near Albuquerque, from encroaching urban development. The issues of tribal sovereignty and cultural preservation have come up against economic interests and political concerns in this and similar disputes, which mirror the larger struggle of the environmental justice movement: to improve the quality of life for the poor and for people of color by empowering women and communities of color.

The New Mexico leaders' environmental beliefs, like their views pertaining to feminism, were, to varying degrees, a function of the history and situation of their racial/ethnic group in the United States, the leaders' motives for political involvement, their identification with their racial/ethnic group, their sense of gender consciousness, and their personal experience of politics. Consequently, the leaders' political goals were also a product of these multiple factors. A principal way of effecting change and achieving one's political goals is through influencing the public policymaking process. We now turn to a discussion of the policy agendas and strategies for policy change employed by the 50 leaders as we examine how they practiced the politics of environmental justice in New Mexico.

CHAPTER 6
Practicing the Politics of Environmental Justice

> I've worked with a lot of grassroots groups over the years. Worked with them in promoting their policies and their programs. Done a lot of that. In the environmental community I've worked with anywhere from the Southwest Organizing Project, which are considered to be a fairly radical group, to other environmental groups who are not that radical. People consider me to be a lot more conservative than what I am. You kind of have to play a role and I hate to say that but it's true. If you want to get something done in the political arena, you have to play within those rules. If you don't play those games, you're not going to last very long. ...Grassroots activism is not for me anymore. I've been coopted by the system. [Laughs] I can be more effective from within the system. You still have to keep pushing, but from a political perspective, I can be more effective from the inside (Jacinta, public official).

UNLIKE OTHER CONTROVERSIAL ISSUES THAT BRIEFLY HOLD THE AMERICAN public's attention and then fade from view, the environment has remained a salient policy concern becoming "one of the most significant issues of our time" (Lester 1989, 1. Also see Dunlap 1989; Downs 1972). This may be due, in part, to the tremendous scope of environmental policy and the fact that its impact is far-reaching; it affects individuals, communities, business and government from the local level to the global. Government regulations are intended to protect consumers' health, to ensure workers' safety, and to minimize environmental degradation in addition to serving other objectives. While political elites play key roles in environmental policy formulation, the public, media, celebrities, educators, and industry have all had an important part in bringing the issues to the forefront of the American political agenda, keeping them there, and increasing support for environmental policy.

Public policymaking, like politics, concerns "who gets what, when, and how." Scholars have long sought to understand how and why individuals organize to effect social, economic and/or political change (see for example Baumgartner and Jones 1993; Sabatier and Jenkins-Smith 1993; Moe 1981; Crenson 1971; Downs 1957). Politics are inseparable from the policy process. The most frequently employed approach within the discipline of political science, the pluralist school, characterizes the American political system as one in which all active and "legitimate" groups can influence the public policymaking process. Whether a particular problem receives the attention of decision-makers, however, depends in large part on whether it has acquired a vocal, organized, and supportive constituency. Therefore, some groups, such as the poor and racial/ethnic minorities, are oftentimes excluded. Hence, the emergence of activists outside of the formal political arena to represent marginalized groups' interests becomes critical to the integrity of the democratic process. As Sapiro points out,

> Political systems are not likely to represent previously unrepresented groups until those groups develop a sense of their own interests and place demands on the system. This requires development of political consciousness and political activism based upon this new group consciousness (1981, 704).

Rational choice theory has gained prominence in recent years as the dominant approach for explaining both individual political behavior and the participation of groups in social/political movements. In his seminal work, *The Logic of Collective Action*, (1965), Olson argued that individuals, in general, will only join groups for the selective incentives or material benefits that are offered. Selective incentives may be either negative or positive (sanctions or rewards), applied selectively to individuals depending on whether or not they contribute to the provision of the collective good. The availability of social selective incentives is limited by the size of the group and by its homogeneity. Larger, heterogeneous groups are more difficult to organize and maintain, Olson believed, because members are less likely to reach consensus regarding the collective good at issue. In small groups, potential differences of opinion can provide an incentive to join, since membership might give the individual a significant influence over policy and the nature of any collective good it would obtain.[1] However, other things being equal, the greater the number of individuals benefiting from a collective good, the smaller the share of gains from participation in the group interest that will accrue to the individual that takes action.

Therefore, lacking selective incentives, motives for collective action decrease as group size increases, resulting in the inability of large groups to act in their common interest, in contrast to small ones.

Scholars have built upon Olson's framework in their attempts to understand more fully the motivation of actors working for political or social change without apparent rewards. For example, Terry Moe (1981) contends that when an individual perceives that she has political efficacy, and believes that her contributions make a difference for political outcomes, she may have every rational incentive to join and contribute to a group. In this case, her decision to participate is shaped by other than economic factors. Instead, she may be motivated by solidary incentives such as friendship or status, or by expressive/purposive incentives deriving from her ideological or moral principles. For example, when one New Mexico leader was asked, "What helps you with your community work?" she responded,

> I think it's the satisfaction when we're successful. Exposing Intel. When parents get a new school... When we participate in voter registration and the turnout increases in that area say from 5 percent to 30 percent... That makes me feel good. When I see someone who comes to us with five kids and is still around four years later. Maybe a 23 year old person with three or four kids who's real disempowered learns how to do some organizing, learns how to use the computer, learns how to speak up and do a press conference, and gets great organizational skills. It's great to see people's personal growth over the years (Salina, activist).

Expressive or purposive incentives may play a role of much greater political consequence than that of selective incentives; they can shape the individual's evaluation of collective goods, and she may also gain a sense of purposive satisfaction from the act of contributing itself.[2] As another New Mexico leader explained,

> Community activism becomes a part of you, becomes dominant in your life—part of my psyche; being a woman in a position of always having to fight. There are some gains but also lots of losses. Feeling oppressed...radicalizes me. As we move into the 21st century and compete for resources it's [increasingly] a challenge. Education, especially, has changed me. I'm who I am because of my involvement (Margarita, activist).

Furthermore, rational choice theory fails to consider the power of group norms and group identity; the social, cultural, and institutional contexts in which individuals make decisions (Tarrow 1991). Just such a context is alluded to by Margarita, above, who felt oppressed by the dominant culture, both as a woman and as an ethnic minority. As one scholar points out:

> Rational choice theory cannot account for the ways group solidarities, moral commitment to the collectivity, and other nonrational values may mobilize people to act independently of individual self-interest. What is rational for the individual is not always consistent with the politically or culturally inspired choices of groups (Eckstein 1989, 4).

Along these lines, Eckstein argues that protest and pressures for change, such as those articulated by the environmental justice movement, are primarily the result of shifting economic relationships and conditions. She adds that the forms of resistance exercised depend on contextual factors: "concerns based on race, ethnicity, and gender, independently and in combination with economic forces, may [also] be sources of disgruntlement that stir defiance" (Eckstein 1989, 4). This is illustrated in the comments of one of the Hispanic leaders.

> When I was working with the Teamsters...I took a group of Latina women out on strike. They worked in a little clothing factory. They had never been told how to fight for themselves, what the rules were, or how the rules applied to them. And afterwards, about five years later, I ran into one of the women I'd been on strike with and she told me she was really glad to see me because everybody wanted me to know that, until I came along, nobody had ever told them how to fight. But now they knew, and so no one was ever going to take advantage of them again, and that made me feel good (Miranda, activist).

Miranda's political activism, like that of many of the New Mexico leaders, was based on her belief that politics is a means for creating greater racial, gender, and social/economic equity, thereby making the world a better place.[3] As Gilkes notes, "[women's] community work comprises both responses to and catalysts for change. Successful or not, it is the effort to make things better and to eliminate the problems and structures that make life difficult" (1994, 238).

Similarly, the environmental justice movement seeks to improve conditions for communities of color by making explicit the connections among racism, sexism, oppression, political disenfranchisement, poverty, and environmental degradation. Indeed, Tesh and Williams see the environmental justice movement as a vehicle for participants to engage in identity politics. Identity politics incorporate

> the lived experiences, common knowledge, and shared values of ordinary people. ...[S]ocial movements [such as the environmental justice

> movement]...represent the collective identities of people who share characteristics, values, or experiences different from the mainstream. ...That social movements also help *create* [a] sense of self is critical to an understanding of their impact on both their adherents and the policymaking process. Social movement ideas and activities inspire previously disparaged or ignored people to think of themselves as a collectivity and to assign political importance to who they are and what they think. They come to feel that their own knowledge, drawn from their own lived experiences is consequential. ...In short, social movements [such as the environmental justice movement] empower their members (1996, 294–295. Italics in the original).

Involvement in the politics of environmental justice enables individuals to effect change through collective action that cuts across gender, racial, class, and geographic boundaries (Di Chiro 1992; Head and Guerrero 1991; Bullard 1990; Carter 1990; Taylor 1990). In contrast to other types of interest groups, environmental group membership, whether in the established national organizations or in the grassroots environmental justice movement, is motivated by ideology, concerns with public policy, and successful "sales" appeals by organizational leadership.

Participation in the environmental justice movement does not necessarily produce any immediate or direct material benefits to the individual, although it may result in the group's (or future generations') access to public goods and/or services. This is illustrated vividly by one of the indigenous New Mexico leaders as she explains why she initially became involved in environmental activism.

> I was interested in water quality issues while I was in law school. I was interested in ensuring that there is a cleaner, safer environment. The reservation should be considered as a reserve. It's kind of a romantic vision... Growing up, there was always a perception that the reservation, it was home. It was clean. It was safe. My interest was based primarily [on preserving] that. And I found out that the tribe has the same status as the states. It's kind of an exercise in tribal sovereignty in that the tribe could be a powerful player in this environmental protection game but they need to be assertive. And this was an opportunity to utilize the federal statutes. And I've always been interested in water issues and in law school I got interested in water law. I could see the potential for really bad environmental degradation (Karen, activist).

The problem of free-riders and of providing material benefits to members as incentives for participation don't affect environmental groups as they do other types of organized interests (Ingram and Mann 1989).

Moe's model for explaining individual participation in social/political change groups is far more applicable than Olson's for understanding the environmental politics and policy agendas of the New Mexico leaders. This is because the women interviewed were motivated primarily by expressive/purposive incentives: their desire to improve the opportunities and quality of life for members of their racial/ethnic communities.[4] (See chapter 4). This is evidenced by the fact that the majority of the grassroots activists employed by their organization earned relatively modest incomes; those not employed by their organization simply volunteered their time. Likewise, both appointed and elected officials had little to gain financially from their political involvement. While New Mexico has a volunteer legislature, many tribal, state, and local positions are also either unpaid or provide only a modest remuneration to elected or appointed officials. Finally, it is common knowledge that highly educated individuals holding law or other advanced degrees (as many of the women interviewed did) can often find higher paying positions in the private sector than in public service. Therefore, as most of these leaders chose careers in public service—in government or in non-profit organizations—it can be assumed that their motivation was not monetary.

GENDER AND PUBLIC POLICY

> Taking women's lives seriously requires that, in political analysis, strategy, and policy recommendations alike, we explore the links between state policy, workplace issues, and community/residential issues (Ackeslberg 1984, 253).

There is considerable documentation that women's political participation, in particular, is motivated by expressive/purposive incentives such as the desire to advocate for and empower others (see for example Hardy-Fanta 1993; Tronto 1993; McCoy 1992; Morgen and Bookman 1988; Darcy et al. 1987; Ferguson 1987; Flammang 1984a; Githens 1983). For example, in a study of urban working-class women, Dabrowski found that

> [t]he women who volunteer claim to be motivated by an altruistic desire to *help* their community. The jobs they perform as volunteers unquestionably aid the general welfare of the area. In the political arena, they are community spokespersons. ...[T]he diversity of their involvement from the schools to service organizations to social action groups, indicates a broad continuum of participation interlocking area social institutions (1983, 432. Italics in original).

Because of women's propensity to work in groups for the greater good of the community, Ackelsberg calls for a model of political participation reflecting women's experience of community-based politics. She describes this as

> a model of social struggle, rooted in and nourished by ongoing social relationships, to replace the pluralist model of interest-group bargaining, a model that masks both the intensity of the feelings and commitments people bring to political life and the power relationships that structure their interactions (Ackelsberg 1988, 308).

As stated in chapter 4, 80 percent of the New Mexico leaders believed that their political involvement was a responsibility that included empowering community members to solve their own problems. Similarly, a growing body of literature confirms that women in political leadership are inclined to advocate for a public policy agenda that promotes equality and improves the status of women (see O'Regan 1997; Kahn 1996; Dodson et al. 1995; Thomas 1994; Tilly and Gurin 1990; S.J. Carroll 1989a; Saint-Germain 1989; Antolini 1984). Moreover, the concerns of the Hispanic and indigenous leaders interviewed for this study are representative of the issues for which other women of color advocate (see for example Acosta-Belén 1993; Hardy-Fanta 1993; Dabrowski 1987; Ackelsberg 1984). Indeed, women of color "find their historical role organized around the nurturance and defense and advancement of an oppressed public family" (Gilkes 1994, 240). These issues of nurturance and care are often ignored and/or overlooked by males. Women leaders, on the other hand, pay serious attention to and prioritize these issues. This is graphically illustrated by one of the New Mexico leaders,

> The people that now are taking the leadership are women. ...The ones with the guts are the women. It's because we're more marginally employed... I also think the bearers of children feel more angered and are more willing to put their necks on the chopping block. In general, women are concerned about the welfare of the future generations. ...One of my main [informers] in the mine was a woman. [She] leaked the information to us that migratory birds were dying from [drinking the water in] the mine's holding ponds. Also, elk and all kinds of other animals. But the men wouldn't say anything (Margarita, activist).

Studies of the political strategies and policy agendas of women activists have revealed gender differences (Naples 1998b; Miller et al. 1996; Women's Environment and Development Organization 1995; Chowdhury and Nelson 1994a; Rodriguez-Trias 1994; Hamilton 1989; Ackelsberg 1984; Boneparth 1984). For example, research indicates that women are more likely than men to favor strongly preservation of the environment (Bellows 1996; Brú-Bistuer 1996; Seager 1996; Thomas-Slayter et al. 1996;

Wastl-Walter 1996; Carpini and Fuchs 1993; Stern et al. 1993; Baldassare and Katz 1992; Di Chiro 1992; Zimmerman 1987; Shapiro and Mahajan 1986; Frankovic 1982). As Rodriguez-Trias simply states, "women's participation and full partnership in discussions on population, development, and the environment are crucial...to crafting policies and programs that will work" (1994, 1379).

Numerous scholars have also examined the effects of gender on agenda setting and the policy process, analyzing differences in the political style, priorities, and influence of female and male actors in formal political institutions (Thomas 1997; Rosenthal 1995a; S.J. Carroll 1992; Kathleen 1989; Antolini 1984; Githens and Prestage 1978). The goals, style, and policy preferences of women in public office tend to be different than those of their male counterparts (see Fox 1997; Rosenthal 1995b; Burrell 1994; S.J. Carroll 1994; Witt et al. 1994). For example, a study of women executives in state government revealed that they were more ideologically liberal than their male counterparts, and that the majority of the women interviewed supported legislation to protect or expand the rights of largely unrepresented groups (female minors, lesbians and gays, and families) (Havens and Healy 1991).

Another study reported that, in general, women legislators introduced more proposals in women's traditional issue areas and on feminist issues than did male lawmakers[5] (Saint-Germain 1989. Also see Kahn 1996; Dodson et al. 1995; Thomas 1994; S.J. Carroll 1991). The author concluded that "proportional group size is an intervening variable in the relationship of gender to public policy in state legislatures" (Saint-Germain 1989, 965) suggesting that for politics to be more democratic, more women need to be elected to office (also see Kahn 1996; Dodson et al. 1995; Burrell 1994; S.J. Carroll 1994). Finally, scholars have found that, due to women's and men's different gendered life experiences, the salience and priority of public policy issues will vary for the sexes, which "can lead to policies that reflect and sustain gendered differences" (Steckenrider and Thomson 1995, 25. Also see Kahn 1996; S.J. Carroll 1991). Such variance in the socio-political attitudes and behaviors of women and men underlines the importance of examining their policy preferences separately.[6]

Differences in the policy choices of women, as compared to those of men, have increased in ways consistent with women's interests and with the agenda of the women's movement. These changes are attributed to women's increased political activity, their greater willingness to express opinions, their higher levels of education, and the overall influence of the women's movement. Most importantly, as women have grown increasingly influential in the policymaking process, their influence in politics has grown (O'Regan 1997; Burrell 1994; Cantor and Bernay 1992; S.J. Carroll 1991; Havens and Healy 1991). Similarly, as women have become increasingly involved in both setting the agenda and in formulating public policy in New Mexico, they have come to exercise greater influence in the state's environmental politics. We now turn to an examination of the public policy agendas of the New Mexico leaders, and the strategies they employed for effecting policy change.

INFLUENCING PUBLIC POLICY, PRACTICING ENVIRONMENTAL POLITICS

The policy agendas of the 50 New Mexico leaders included a broad array of concerns ranging from land use planning to the provision of educational programs. Although the scope of their work extended from solving neighborhood level problems to forming networks with organizations internationally, their common focus was achieving environmental justice for their racial/ethnic communities. This entailed providing for the health and safety of community members, empowering politically marginalized individuals and groups, and establishing programs to meet their social and economic needs. As scholars note,

> [P]ublic action on environmental issues generally produces mixed results regarding equity. Increasingly, environment and equity concerns cannot be treated separately (Paehlke and Rosenau 1993, 674).

Consequently, Bullard's often-cited studies of the environmental justice movement indicate that the concerns of movement leaders largely arise from environmental and economic conditions (Bullard 1995, 1993, 1990; see also Rocheleau et al. 1996; Sen 1995; Rodriguez-Trias 1994; Di Chiro 1992; Taylor 1989). Another researcher has discovered that Mexican American community organizers pursue two distinct agendas in their struggle for environmental justice: (1) general social equity, including long standing problems like poverty and unemployment and (2) the right to participate as equal partners in environmental policy decision-making (Marquez 1994).

Additionally, studies have found women to be more supportive than men of government initiatives relative to public health, employment, reducing the income gap between rich and poor, social programs, and the decommissioning of nuclear power plants (Carpini and Fuchs 1993). Furthermore, Krauss notes that for women in the environmental justice movement, politics is tied largely to their lived experience and to their identity as mothers, as women of color, and as members of their particular ethnic community.[7]

> Unlike the more abstract, issue-oriented focus of national groups, [these] women's focus is on environmental issues that grow out of their

> concrete, immediate experiences... What emerges is an environmental discourse that is mediated by subjective experiences and interpretations and rooted in the political truths women construct out of their identities as housewives, mothers, and members of communities and racial and ethnic groups. ...Through their informal networks, they compare notes and experiences and develop an oppositional knowledge used to resist the dominant knowledge of experts and the decisions of government and corporate officials (Krauss quoted in Tesh and Williams 1996, 295–296).

My findings support all of the above observations, which are generally applicable to both the grassroots activists and the elected and appointed officials interviewed.

The public policy agendas of the New Mexico leaders were comprised of 227 individual policy initiatives, which, in turn, were grouped within three broad categories. These consisted of policies or programs that promoted (1) human rights, empowerment, and self-determination; (2) community planning and development; and (3) conservation of natural resources and environmental quality.

Table 20. Leaders' Policy Initiatives

	Activists		Officials			
Policy Area	Indig.	Hisp.	Indig.	Hisp.	Total[1]	%[2]
Human Rights, Empowerment, and Self-Determination	27	27	36	18	108	48
Community Planning and Development	6	3	16	43	68	30
Natural Resources and Environmental Quality	17	21	4	9	51	22

[1] Refers to the number of policy initiatives promoted by each group of leaders. Some leaders did not support any of the initiatives within a certain policy area while other leaders advocated several initiatives within each policy area.

[2] Refers to the percentage of policy initiatives in each policy area out of a total of 227 individual initiatives.

The first category (human rights, empowerment, and self-determination), which included goals such as cultural preservation, assertion of individual and group rights, and increasing participation in public decision-making, received overwhelming support from the New Mexico leaders with 108 separate initiatives. Nearly all of the women interviewed (47 or 94 percent) included policies of this kind in their political agendas. Approximately half of the leaders (26 or 52 percent) sought to benefit their racial/ethnic communities by advocating 68 policies in the second category (community planning and development), which provided greater education and employment opportunities and addressed environmental, public health, and safety concerns. Finally, a majority of the leaders (30 or 60 percent) reported having some sort of environmental policy goal, which amounted to 51 separate initiatives to maintain or improve water, air, or soil quality. We will now examine each of these policy categories in greater detail.

Human Rights, Empowerment, and Self-Determination

The vast majority of the leaders (94 percent) promoted policies relating to human rights and self-determination. As we can see from Table 21, in all, 108 such policies were articulated by the New Mexico leaders.

Table 21. Leaders' Policy Agendas Promoting Human Rights and Self-Determination

	Activists		Officials			
Policy Area	Indig.	Hisp.	Indig.	Hisp.	Total	%
Individual and Group Rights[1]	2	5	19	11	37	74
Indiv./Comm. Empowerment[2]	9	12	7	7	35	70
Cultural Preserv./Repatriation[3]	9	10	4	0	23	46
Tribal Sovereignty	7	0	6	0	13	50[4]

[1] Refers to the rights and services due to particular populations including: children, families, seniors, women, and workers.
[2] Refers to the empowerment of individuals and communities of color.
[3] Refers to the preservation of cultural practices, to the repatriation of native lands and artifacts, and to the return of Spanish land grants to Spanish descendants.
[4] This figure represents the percentage of indigenous leaders who included tribal sovereignty among their policy goals.

While support for social programs was also mentioned, human rights initiatives were generally combined with policies rendering related social services to particular populations. For example, proponents of children's rights, families' rights, and/or women's rights simultaneously advocated the provision of public services to meet the particular needs of each of these groups. Workers' rights and safety and services benefiting senior citizens were also included. As one activist stated,

> [our goal is] to organize the unorganized, to empower the workers to represent themselves. To educate them to understand their rights. To educate them to defend themselves in contract or workplace violations; there are lots of health and safety issues out there (Esperanza, activist).

In all, 37 policies were mentioned relating to the rights of individuals and to particular populations and to their attaining public services.

Of the policies proposed by the leaders in this category (human rights and self-determination), a total of 35 involved facilitating the empowerment or self-determination of individuals and their racial/ethnic group. These women tended to view politics

> as the web of activities in which people engage out of concerns generated by their daily lives. Politics is about attending to the quality of life in households, communities, and workplaces. Thus politics becomes the positive empowerment of citizens through democratic participation on a variety of levels (Longo 1997, 12).

Leaders advocating empowerment often sought to reform government and other organizations by increasing public participation and transforming policymaking bodies so that they would be more receptive to citizens' needs. The ultimate goal of these particular New Mexico leaders was to make public organizations more inclusive by "opening them up" to greater citizen involvement in order to better reflect the interests of different racial/ethnic groups, and of women, generally. For example, as two public officials stated, "I want to see women move up" (Gina), and "[I want] to make information easily accessible to the public" (Estela).

Specifically, 35 policies contributed to the empowerment of individuals and/or their communities; 23 initiatives related to the preservation of cultural practices, to the repatriation of native lands and artifacts, or to the return of Spanish land grants to family heirs; and 13 policies (all advocated by indigenous leaders) concerned the exercise of tribal sovereignty. One indigenous leader whose activism involved cultural preservation, repatriation of native peoples' lands and artifacts, and litigation to maintain tribal sovereignty described her work in the following way.

> Symbolically and figuratively, we collect the words that remain of our language, our culture, our practices. We collect the artifacts...the concepts of our religion. ...We are the gatherers (Linda, activist).

Interestingly, while seven of the policies relating to individual and group rights were advocated by grassroots activists, 30 of these initiatives were supported by public officials. In particular, public officials sponsored legislation relating to domestic violence, they increased the awareness of tribal councils to the problem of spousal- and child abuse; they crafted and/or implemented the Indian Child Welfare Act as well as state and tribal legislation pertaining to services for children and families; and they developed legislation to improve social services to families, single mothers, and children. Furthermore, both elected and appointed officials worked to educate others about women's issues; to increase women's input in tribal and other governmental policymaking processes; to increase racial/ethnic minority and women's participation in public commissions and other decision-making bodies; to amend their tribal Constitution to allow for greater participation in tribal decision-making; to develop and lobby for the passage of the American Indian Religious Freedom Act and the Indian Mineral Development Act as well as other legislation affecting indigenous people; and they worked to ensure that the U.S. government fulfilled its treaty obligations to tribes.

On the other hand, nearly twice as many policies relating to empowerment, cultural preservation, and sovereignty (47 versus 24) were advocated by activists than by officials. As one leader explained, her policy goals included

> help[ing to] create real democratic structures, empowering communities, using education. To get community residents to develop a global perspective of issues, since it's all interconnected. They pick up skills through the work we do (Salina, activist).

Part of the difference between the policies supported by the officials and activists may be explained by their peculiar political roles. The granting of rights and the provision of public services are both (to different extents) within the purview of elected or appointed officials. Public officials, whether in the legislative, executive, or judicial branches, are more likely than grassroots activists to have the power to establish, fund, or oversee the administration of public services. Conversely, activists, in contrast to public officials, are relatively free to advocate for more extreme policy change and to represent the interests of politically marginalized groups such as the poor, children, and communities of color.

Similarly, greater numbers of both (1) human rights policies and (2) policies relative to empowerment, cultural preservation, and sovereignty were promoted by indigenous leaders (21 and 42, respectively) than by Hispanas (16 and 29, respectively). Although both American Indians and Hispanics in this country tend to experience higher levels of poverty than other racial/ethnic groups, the social and economic problems endured by native people, especially by those living on reservations, have been particularly severe, pervasive, and intractable. Furthermore, American Indians have a unique relationship with the federal government through which native peoples' rights, and their access to services, are guaranteed in treaties with the United States. However, more often than not, these treaties have been violated, as have been the rights of indigenous peoples. As a result, native leaders must be vigilant, guarding their tribal rights and pressing for the legally mandated provision of services to their communities. One tribal official explained that the programs she developed while in office "will benefit the tribe, the people" (Tracie); another Indian leader said, "I speak for people who can't speak for themselves" (Tanya, public official).

Community Planning and Development

Community planning and development issues encompassed economic development; opportunities for quality educational programs; provision of affordable quality health care and maintenance of a healthy environment; and a bundle of quality of life issues including: provision of public transit, reduction of crime and violence, provision of affordable housing, and regulation of the insurance industry for the protection of consumer rights.

Table 22. Leaders' Policy Agendas Promoting Community Planning and Development

	Activists		Officials			
Policy Area	Indig.	Hisp.	Indig.	Hisp.	Total	%
Economic Development	0	2	2	3	7	14
Education: Opportunity/Quality	2	0	6	4	12	24
Health Care / Environt. Health	3	1	5	4	13	26
Quality of Life Issues	1	0	3	7	11	22

As was the case with the provision of human rights discussed above, community planning and development concerns were cited more frequently by public officials than by grassroots activists. This was probably for the same reasons, as elected and appointed officials are in a better position than community organizers to create, fund, or implement such initiatives. For example, some of the specific concerns articulated by the public officials included establishment of bilingual education programs in New Mexico's public schools—not for Spanish speakers as these programs are fairly well established, but for Navajo children wishing to maintain their language;[8] development of insurance industry regulations to protect the health of mothers and their newborn infants as well as protecting the health of Medicaid recipients; improvement of public transit systems in the state's urban centers; and the establishment of honors programs in Albuquerque public high schools with large minority student populations. Where grassroots activists did promote development policies, these were often tied to maintaining cultural preservation and to providing communities of color greater means of self-sufficiency. This is illustrated by Marta's comments about the outcome of a community development program.

> [I hope] that the emerging leaders...acquire depth in management skills, in visioning, in being able to develop sustainable businesses, and in doing program development. ...Our approach...is to demonstrate what people can do when they decide to take their own culture and economic system in their hands and deal with the issues of poverty and hopelessness and cultural alienation (Marta, activist).

Sustainability, which refers to the "long-term relationship between our industrial economy and the natural resource base that sustains it," is particularly important for economically disadvantaged groups because they "spend a disproportionate share of their incomes on both energy and food" and feel the impact of price increases more than do affluent consumers (Paehlke and Rosenau 1993, 678).

Policies relating to housing, crime, and/or public transit were mentioned ten times: nine by public officials and one by a grassroots activist. However, nearly even numbers of indigenous women (4) and Hispanas (6) supported these initiatives. The greatest number of policies in this category (19) had to do with economic development and education. Again, while more officials (15) than activists (4) included such policies in their political agendas, nearly even numbers of native women (10) and Hispanas (9) advocated these issues. Finally, health care provision, environmental health, and regulation of the insurance industry were policy items mentioned by more than twice as many officials (10) than activists (4) although, as before, support for these issues was nearly evenly split between indigenous (8) and Hispanic (6) leaders. When activists did advocate for issues of health and safety, for example, they were frequently couched in environmental terms and linked to broader democratic concerns for community participation and government accountability. For example, one leader explained that her organizational mission was "to influence the powers that be. To encourage the citizenry to ask questions, to demand safety, to demand honesty in dealings with the [U.S.] Departments of Energy and Defense, and the state about issues [regarding] radioactivity" (Susana, activist). Similarly, as a victim of toxic workplace conditions, another activist fought

> to see a hospital [built] for all these people exposed to chemicals [in their workplace]. ...To one day have a hospital where we can all go and be seen by professional people who know what's going on in our lives, what's going on in our minds, why we feel this way. And the anger we have within ourselves (Monica, activist).

Public health, economic development, and environmental quality were closely linked, as illustrated by the experience of Lois Gibbs. The housewife-turned-activist is a well-known environmental leader who was politicized by personal tragedy. Her involvement in exposing the health hazards of Love Canal, a severely contaminated residential subdivision, mobilized her to join the growing movement against the production and illegal dumping of toxic wastes. She subsequently established and currently heads the Citizen's Clearinghouse for Hazardous Wastes, a national organization for consumer advocacy. Issues such as public health, economic development, crime, social services, and consumer rights all fall within the scope of the environmental justice movements' agenda because they contribute to promoting and maintaining the community's physical, economic, and social well-being. We now turn to the last category of policies and programs, that which deals primarily with environmental concerns.

Conservation of Natural Resources and Environmental Quality

As stated earlier, 60 percent of the New Mexico leaders advocated environmental policy agendas,[9] which comprised one or more of the following issues: water quality and/or water rights (12 or 24 percent); conservation of indigenous or Hispanic peoples' traditional lands and land uses, and long-range planning for land use in urban areas (21 or 42 percent) including preservation of soil quality and waste management; and general environmental protection (17 or 35 percent) including production and use of renewable energy sources. (See Table 23, below).

Several of the leaders interviewed believed that communities of color are specifically targeted for placement of polluting industries and facilities. To support their claims, they cited specific examples of cancer clusters, the contamination of well sites, incidences of health and safety violations by both government and industry, and other such problems. Their efforts to publicize these concerns resulted in the production of video documentaries aired by local public television stations, the publication of numerous studies and position papers by various organizations, public hearings held around the state, and feature stories in the news media.

> ...We write letters [to get on the agenda] whenever they have EPA hearings. I've testified at three. When the EPA comes down from Washington, you know, you stand up and tell them whatever you think they're doing wrong. They listen very nicely and then go out and do precisely what they wish. [Laughs] We get a lot of the information and we have resident experts, or people we can count on to translate what they really mean to the lay people. Then they help us prepare our objections and make them respond to us. Our main hope is that [WIPP][10] would not get a RCRA permit.[11] ...The head of the state environmental department seems to be working with us, affording us the opportunity to go to these meetings. Most of our strategies revolve around basically reviewing their documents and demanding to know what's going on, to have them explain it, to bring it out. We can go in

> and say "Hey, we've seen studies by doctors saying that this is a dangerous thing and you're not proceeding" because, my God, they are incorrigible. [Shakes her head] If you don't watch them every minute they will try to sneak something [over] on us (Susana, activist).

While 51 individual policy directives were cited, Hispanas included a majority (59 percent or 30) of the environmental policies in their agendas. In contrast, fewer than half of these policies (41 percent or 21) belonged to the indigenous leaders' agendas.

Table 23. Leaders' Policy Agendas Promoting Conservation of Natural Resources and Environmental Quality

	Activists		Officials			
Policy Area	Indig.	Hisp.	Indig.	Hisp.	Total	%
Environmental Protection[1]	2	10	2	4	18	36
Land Conserv./ Land Use[2]	9	9	1	2	21	42
Water Quality / Water Rights[3]	6	2	1	3	12	24

[1] Includes adoption of regulations and/or implementation of programs to protect water, air, and soil quality. Also includes the production and use of renewable energy sources.
[2] Refers to both the conservation of indigenous or Hispanic peoples' traditional lands and land uses, and to long-range planning for land uses in urban areas. Also includes (1) adoption of regulations and/or implementation of programs to reduce soil erosion and to preserve arable lands, and (2) implementation of programs for waste reduction, collection, and disposal.
[3] Includes adoption of regulations and/or implementation of programs to improve or preserve water quality, and to maintain access to water supplies.

The greater support of environmental policies by the Hispana leaders may be a result of the prioritization of social services and human rights as more pressing concerns for native women. On the other hand, it is important not to make generalizations about the salience of environmental issues to indigenous peoples based solely on these findings. To the contrary, American Indian nations have been disproportionately impacted by various types of projects that have destroyed significantly the natural environment. These have included "the strip-mining of coal, oil and gas pipelines, power dams, uranium mines, and oil shale development" as well as proposals to place storage facilities for toxic and/or radioactive waste on Indian lands (Paehlke and Rosenau 1993, 683. Also see Diane 1995; Fleck 1995; Austin and Schill 1994; Jones 1994; Gedicks 1993; House 1993; Churchill and LaDuke 1992; Ambler 1991; Taliman 1991b). As a result, numerous native activists have emerged as leaders in the environmental justice movement (see Mack-Canty 1997; Laws 1995; Sen 1995; Conrad 1991; LaDuke 1991; Taliman 1991c; Redhouse 1984).

Most of the environmental initiatives (75 percent or 38) made by the New Mexico leaders were advocated by grassroots activists; only about one quarter (25 percent or 13) were policies supported by public officials. For example, the activists pressed for policies that would: provide information and technical assistance to communities of color for the purpose of developing environmental programs; facilitate the negotiation of agreements between communities and industry or government agencies for pollution abatement and site clean-up; encourage increased citizen participation in environmental policymaking; create programs for the prevention of soil and groundwater contamination; and provide reparation to victims of toxic poisoning. The policies mentioned by officials included similar components such as the development of regulations and implementation of programs to enhance air and water quality; the clean-up of toxic sites; ensuring tribal control of mineral rights on Indian lands; and the formulation of land use plans to guide urban development. As one leader stated,

> For several years, I've sponsored a beverage container recycling act—the "bottle bill." I carry legislation that relates to sewer and water hook-ups for low income people so that we can get them into systems that will give them safer water. I like issues that relate to landfills; reduce and reuse of materials; safety for our water table. Mostly anything that protects our water and air; our infrastructure (Sonia, public official).

As is characteristic of the environmental justice movement, the leaders' policy agendas overlapped among the three categories so that human rights and public service initiatives were often tied to traditional environmental concerns, which were frequently linked to development issues. For example, promoting sustainable economic development, preserving traditional agricultural practices, and improving the Rio Grande's water quality were equally important goals for one leader who stated that her organization's mission was "to restore the water to drinkable quality and to keep the communities active and alive along the Rio Grande" (Miranda, activist).

Activists, in contrast to elected or appointed officials, could promote a narrow policy agenda while determining program priorities for their particular organization. Elected officials, on the other hand, had to answer to a larger constituency representing a broad range of (possibly conflicting) policy preferences. The prioritization of policies and programs by elected officials depended on numerous factors such as issue salience, the availability of public funds, whether they had the support of the administration and/or special interest groups, media coverage, the general public sentiment, and so on. Similarly, appointed officials had to follow the policy lead of the administration responsible for their appointment. Furthermore, the legal mandate of their board, commission, or agency often limited the scope of their political influence to a specifically designated issue area and clientele. In some cases, the nature of a leader's public office or the specific mandate of her organization did not allow her to advocate for particular groups. Rather, her responsibility was to a broader constituency of, for example, consumers, children, or students. To put it simply, with regard to their policy agendas and program priorities, public officials were far more constrained by the nature of their position and by electoral politics than were the grassroots activists.

Regardless of whether they were activists or officials, however, preparing community members to effectively participate in the political decision-making process was an important component of each leader's policy agenda. The New Mexico leaders promoted consensus building, shared decision-making, and community participation in goal-setting and public policy formulation. Each leader exhibited egalitarian views and supported participatory forms of democracy. Some strove for political and economic transformation of their racial/ethnic community by increasing public awareness and access to information through forums, workshops, and public hearings. Others sought to empower community members through leadership training and/or education. In several cases, grassroots activists recruited their membership to promote gender equality and ethnic/racial diversity and to ensure representation of poor and low-income citizens in their organizations' leadership. In general, the New Mexico leaders sought to increase political participation by empowering community members to help themselves, by providing advocacy, by facilitating coalition building, and by sharing information. To accomplish these goals, and to influence the public policymaking process, they employed a multitude of strategies.

STRATEGIES FOR EFFECTING POLICY CHANGE

The strategies employed by the New Mexico leaders generally fell under one of five headings: (1) strategies to educate and inform, (2) community relations strategies, (3) partnership-building strategies, (4) strategies to develop resources, and (5) strategies using the existing legal and political systems.[12] The individual strategies contained within each of these five groupings can be further classified as "conventional," "unconventional," or "mutually employed," as numerous techniques were commonly used by both political "insiders" (the public officials) and political "outsiders" (the grassroots activists). A brief discussion of these classifications follows a description of the strategies.

Strategies to educate and inform often incorporated capacity building: establishing leadership training for community members to increase public participation in policymaking; conducting workshops, training seminars, or conferences to educate and inform; educating and mentoring youth; producing publications such as newsletters and brochures for wide distribution; and conducting voter registration drives or sponsoring political debates. As one grassroots activist explained,

> If people have information, there's an enormous amount of energy and creativity and intelligence in the community. All that had to happen was for people to get the information and find out the methods by which to use that information [so they could] participate as citizens. ... No one is going to fight our battles and protect our interests but ourselves (Juana, activist).

Community relations strategies included gaining advocacy, endorsement, or sponsorship from politicians or celebrities; producing audio-visual programs for public presentation; holding well-publicized hearings to inform and garner public support; effectively using free media such as radio, television, or newspapers to garner public support; and organizing rallies, demonstrations, boycotts, or strikes. Several of these strategies were successfully employed by one Pueblo's leadership in their efforts to secure passage and enforcement of the tribe's own environmental regulations.

> We...started a letter writing campaign and it took almost a year, but we were able to get William Reilly, the head of EPA out here. And we hosted the meeting at [the Pueblo]. The purpose of that was to get him on the reservation, face-to-face with the Indians because he'd never been. It was the first time he'd ever been on a reservation, much less in a council chamber full of Indians. So it was really interesting. As a matter of fact, the next day our water quality standards were approved and we got our statement of state status. It was pretty neat, because nobody knew who he was, right? People didn't fathom that he was the Secretary of the EPA. [Laughs] He was just a white man as far as everybody was concerned, so everybody was real informal and relaxed (Valerie, public official).

Valerie's strategies included holding workshops to educate the tribal administration on environmental law and public hearings to inform the community on the need for water quality standards; gaining the endorsement and assistance of an influential federal official—the head of the U.S. Environmental Protection Agency; acquiring technical assistance and expertise about environmental safety regulations; and lobbying the City of Albuquerque, the New Mexico Environment Department, and state legislators to support the tribe's establishment of their own water quality standards.

Partnership strategies involved forming alliances with business and industry to invest in communities of color; networking to exchange information, to share resources, or to develop strategies; and building coalitions with civic, social, labor, or professional organizations, religious groups, or governmental agencies.

Strategies to develop resources included developing reliable, credible sources of printed information based on accurate research; raising funds through grant-writing or other means to gain financial support for operations and projects; repatriating land, cultural and/or religious artifacts for preservation of community heritage; providing or acquiring technical assistance or expertise; and employing technical or legal experts as witnesses, advocates, or spokespersons. Jacinta knew that her agency would face opposition when they sought passage of tougher state environmental standards. Therefore, she and her staff developed a plan for increasing public awareness and broad-based support for the new legislation by using various educational, community relations, and partnership-building strategies.

> We got the new [state environmental] regulations in '91 written, passed and through, and we got the ... plan passed in '92. We were real effective in putting together teams of constituencies. We didn't make decisions just based on what one group or another wanted. We did a lot of constituency negotiations between environmentalists and industry groups and business people and other public interest groups. So we had input from various entities that were affected by the policies that we had, and the decisions and regulations that we made. And that worked out really well (Jacinta, public official).

Strategies using the existing legal and political systems incorporated the following: reviewing official documents and/or monitoring regulatory enforcement and compliance; lobbying to affect legislation at the local, state, tribal, or national level; litigating to enforce compliance with laws or to win rewards for damages; and influencing local, state, tribal, and/or national political or bureaucratic officials.

While almost any of these techniques for affecting public policymaking can be utilized by either public officials (political "insiders") or grassroots activists (political "outsiders"), some strategies are more appropriate for one set of leaders than the other. To illustrate, organizing a protest march to raise public awareness of an issue is more likely to be a technique used by "outsiders" who lack the influence and ease of access that "insiders" have to affect the policy process. In this case, political "outsiders" use an "unconventional" strategy to get their concerns placed on the public policy agenda. On the other hand, "insiders" such as elected and appointed officials are more likely to hold formal public hearings, for example, to gather input on or to "sell" a policy initiative to community residents. Because the "insiders" control the political forum and set the public policy agenda, they don't need to resort to "unconventional" strategies.

"Unconventional" or "informal" political strategies incorporate such activities as leafleting, holding non-partisan public forums and workshops for community residents, performing plays, holding "sit-ins" or "teach-ins," and organizing boycotts as well as engaging in direct action tactics like demonstrating, picketing, and occupying public spaces or buildings. In contrast, "conventional" or "formal" political strategies generally include activities such as voting, petitioning, campaigning, fundraising, registering voters, running for public office, and lobbying policymakers. These are the traditional sorts of activities we tend to associate with the practice of electoral politics.

These categories are not hard and fast. Rather, over time, American society's conceptions of "unconventional" political behavior changes. The civil rights movement, for example, was instrumental in establishing the legitimacy of "sit-ins" and marches. During the late 1960s and the 1970s, many mainstream citizens participated in these forms of political participation. Leaders of grassroots organizations such as the Sierra Club and Public Citizen (Ralph Nader's consumer advocacy group) once relied on unconventional techniques like demonstrations or boycotts. However, as these actors and their groups gained political influence and legitimacy, they increasingly adopted conventional techniques for affecting policy change. Moreover, many of the strategies that they employed such as litigation are becoming "conventional" or at least widely used by a variety of political actors. My point here is that the way in which political activities are classified is subject to change. For purposes of this research, however, I have classified the strategies used by the New Mexico leaders in the following ways.

"Unconventional" strategies include:

- providing leadership training for community members
- conducting workshops, training seminars, or conferences
- organizing rallies, demonstrations, boycotts, or strikes
- coalition-building
- repatriating land or cultural or religious artifacts
- lobbying bureaucratic officials (in contrast to elected or appointed officials)
- litigating to enforce compliance with laws.

"Conventional" strategies include:

- conducting voter registration drives or sponsoring political debates
- monitoring enforcement and compliance of policies, programs, or laws

- lobbying public officials (elected or appointed officials)
- campaigning for a political candidate or contributing to their campaign
- running for public office.

In addition to these so-called "unconventional" and "conventional" techniques for affecting policy change, there exist numerous strategies appropriate for use by both political "insiders" and "outsiders." Of the strategies used by the New Mexico leaders, those employed by both public officials and grassroots activists include:

- educating and mentoring youth
- producing publications for wide distribution
- gaining advocacy, endorsement, or sponsorship from politicians or celebrities
- producing audio-visual programs for public presentation
- holding public hearings
- using free media such as radio, television, or newspapers
- forming alliances with business and industry
- networking
- developing research reports
- fundraising
- providing or acquiring technical assistance or expertise
- utilizing technical or legal experts as advocates.

In most cases, the leaders used a variety of strategies – whether "unconventional," "conventional," or common to both "insiders" and "outsiders" – depending on the situation and their perceived effectiveness. As one activist explained,

> We try to get people to attend hearings with us. We write grants, letters. We try to [get bills passed] in the legislature that are beneficial to us, to the people of New Mexico. We have fundraisers. We spend a lot of time gathering information. We do a lot of proselytizing, if possible. We do some administrative tasks, office keeping. We put out a newsletter three or four times a year...I don't like the administrative work, [grinning] I prefer the "in your face" type stuff (Susana, activist).

Table 24, below, provides a breakdown of the mean, number, and types of strategies from each of the five categories utilized by the New Mexico leaders. Table 25, which follows, similarly displays the mean, number, and classification of the strategies used.

Table 24. Types of Strategies Used by Leaders to Influence Public Policymaking

	Activists		Officials		
Type of Strategy	Indig.	Hisp.	Indig.	Hisp.	Total
Educate and Inform	3.2 (41)[1]	3.5 (45)	1.4 (18)	1.7 (19)	2.5 (123)[2]
Community Relations	1.7 (22)	2.8 (37)	1.4 (18)	2.0 (22)	2.0 (99)
Partnership Building	1.8 (23)	2.3 (30)	2.3 (30)	2.2 (24)	2.1 (107)
Resource Development	3.6 (47)	3.6 (47)	3.9 (51)	2.6 (29)	3.5 (174)
Use Existing Systems	2.6 (34)	3.4 (44)	3.3 (43)	3.0 (33)	3.1 (154)
Total	12.9 (167)[3]	15.6 (203)	12.3 (160)	11.5 (127)	13.2 (657)[4]

[1] Mean and total number of strategies in this category employed by a particular group of leaders.

[2] Mean and total number of strategies in this category employed by all 50 leaders.

[3] Sum of the mean and total number of strategies in all five categories employed by a particular group of leaders.

[4] Sum of the mean and total number of strategies in all five categories employed by all 50 leaders.

Table 24 indicates that the New Mexico leaders reported using a total of 657 strategies with an average of 13.2 employed by each leader. Roughly, the leaders used from one to three techniques from each of the five types of strategies. There was very little difference with regard to the average number of strategies used from each category. For example, the most popular were resource development strategies with an average of 3.5 techniques used per leader; yet even the least popular, which were community relations strategies, averaged two per leader.

While Hispanic officials employed the least number of strategies (11.5) from all five categories, Hispanic activists used the most (15.6). The difference between the average number of strategies utilized by Hispanic leaders (13.8) and the average for indigenous leaders (12.6) was slight. As a whole, the grassroots activists used a slightly greater variety of techniques for influencing policymaking (on average 14.2), than did the public officials (12). This was expected because, in contrast to leaders in formal electoral politics, grassroots activists

> [in] autonomous, less traditionally structured groups have the advantage of innovation, developing work styles especially suited to their goals such as collective decision-making and engaging in protest activities without risking hard-earned political legitimacy (Boneparth 1984, 282).

In other words, because grassroots activists face fewer constraints on their political behaviors than do public officials, they can employ a greater variety of "unconventional" strategies to influence change.

Table 25. Classification of Strategies Used by Leaders

	Activists		Officials		
Strategy Used	Indig.	Hisp.	Indig.	Hisp.	Total
"Unconventional"	3.8 (49)[1]	4.4 (57)	2.4 (31)	2.0 (22)	3.2 (159)[2]
"Conventional"	2.3 (30)	2.8 (37)	2.8 (37)	2.8 (31)	2.7 (135)
Mutually Employed[3]	6.8 (88)	8.4 (109)	7.1 (92)	6.7 (74)	7.3 (363)
Total	12.9 (167)[4]	15.6 (203)	12.3 (160)	11.5 (127)	13.1 (657)[5]

[1]Mean and (total) number of strategies in this class employed by a particular group of leaders.
[2]Mean and (total) number of strategies in this class employed by all 50 leaders.
[3]"Mutually Employed" strategies are comprised of either unconventional or conventional strategies that are used by both activists and public officials.
[4]Sum of the means and (total) number of strategies in all three classes employed by a particular group of leaders.
[5]Sum of the means and (total) number of strategies in all three classes employed by all 50 leaders.

Of the 657 strategies used by the New Mexico leaders, we can classify 159 (24 percent) as "unconventional," 135 (21 percent) as "conventional," and the majority, 363 (55 percent), as "mutually employed" since these were techniques commonly used by both the activists and officials. While the Hispanic activists each used on average 4.4 "unconventional" strategies, this number was not much higher than the average quantity (two) employed by the Hispanic officials who used these particular techniques the least. Nearly all the leaders, whether activists or officials, reported using an average of two (2.3 to 2.8) "conventional" techniques to affect policy change. Overall, the leaders used an average of three (3.2) "unconventional" techniques, nearly three (2.7) "conventional" techniques, and seven (7.3) mutually employed techniques. Clearly, there was not a substantial difference between the use of "unconventional" and "conventional" techniques by the grassroots activists and public officials.

Based on the long-held assumption in political science that significant differences exist between grassroots activists and public officials, few studies have compared just how politics is practiced by these two groups of actors. Instead, we have taken for granted that their public policy agendas contrast substantially, and that "conventional" strategies are employed by actors in electoral politics while those used by grassroots activists are "unconventional." However, there was actually very little variation between the New Mexico grassroots activists and public officials with regard to either the quantity or different sorts of strategies used for influencing public policy. To illustrate, while Burrell suggests that women legislators impact public policymaking by "affecting the operational process of the system, influencing policy outcomes, and reshaping the agenda to include women's concerns" (1994, 153), these strategies were employed by both the New Mexico grassroots leaders and public officials.

Boneparth (1984) suggests that women in elected and appointed office and women in community-based organizing pursue a "combined insider-outsider strategy" by coordinating their efforts to more effectively influence public policy. This was certainly the case for several of the New Mexico leaders, especially as many of the elected and appointed officials who had been grassroots activists maintained their long time connections with their former community organizations. For example, two of the Hispanic leaders who held positions in municipal government remained involved with the neighborhood associations that they had helped to found years earlier. Similarly, other former activists who had entered electoral politics built on their experience and connections with grassroots organizations to advocate for particular programs and/or legislation.

Many of the 50 leaders who new each other frequently networked and cooperated on projects or campaigns of mutual interest by sharing information, pooling resources, or using their influence to further a common goal. While most of the cooperation tended to occur among women belonging to the same groups – for example among native tribal leaders or among Hispanic community activists – their alliances often crossed such boundaries. To illustrate, both native and Hispanic activists provided technical and/or legal assistnce through their grassroots organizations to tribal governments facing environmental and public health problems. Leaders of one particular grassroots environmental group assisted and/or formed coalitions with five other grassroots leaders participating in this study. Leaders of the environmental group also worked with at least two of the public officials interviewed by organizing forums and public hearings, and by participating in negotiations with polluters, community members, and government agencies. Where possible and advantageous to do so, the New Mexico leaders effectively combined "insider-outsider strategies" to further their policy goals.

These findings are significant in that no major distinctions existed among the variety or types of strategies employed by the leaders. The fact that the activists and officials similarly pursued pretty much the same broad range of techniques for influencing public policy reveals greater convergence than divergence on their agendas and strategies for change. In fact, the New Mexico leaders employed a wide range of strategies, which simultaneously promoted and advanced their particular policy agendas. By this, I mean that the techniques they used to promote the policy of making politics more inclusive, for example, actually involved empowering community members by giving them organizing and public speaking skills, by teaching them to locate and effectively utilize resources, by increasing participation in public decision-making processes, and by mentoring youth to assume tomorrow's leadership positions.

It's not surprising that both the grassroots activists and public officials promoted policies to increase community participation in public decision-making while they utilized strategies for policy change that incorporated empowerment and consciousness-raising activities. Indeed, both this agenda and the techniques employed for empowering others seem to be characteristic of many women in positions of political leadership, whether they are community activists or elected or appointed officials. As one study of leaders in state government found, nearly all of the female executives interviewed stated that

> women leaders have reshaped the policy agenda, bringing new issues into the public policy arena...includ[ing] child care, family and children's issues, equal employment policies and other work issues with special effect on women...In addition, [to] welfare reform, consumer protection, privacy regulations...and civil rights... Those interviewed also saw a trend in departments with women in leadership to approach policy analysis, development and implementation with sensitivity for the potential impact on women workers and women clients or constituents (Havens and Healy 1991, 65).[13]

These findings are particularly important when one considers that "the entire policy process...is shaped by a social fabric in which hierarchies based on gender, class and race are heavily intertwined" (Diamond and Hartsock 1981, 720) and which tends to exclude the voices of women and racial/ethnic minority groups: those who are less likely to exert political influence or economic power in society. Furthermore, in a study that examined citizen input into national environmental policymaking, Zeitler noted that

> [c]ontemporary environmental policies often reflect the input of vested business, development, and industry interests. Bureaucratic, scientific, legal, and economic "experts" representing private interests control the terms of environmental debates, thereby marginalizing public input in the policy process. This marginalization hinders the realization of "environmental democracy" within American political culture and distinctly shapes environmental policy outcomes (1996, 1).

The process of formulating public policy requires that government provides the means by which residents, both as individuals and as representatives of groups, are enabled to participate actively. The New Mexico leaders recognized that community members must be included as principal actors in the policymaking process to identify local needs, to permit consultation prior to policy design, during program implementation, and throughout evaluation procedures. We can infer from the extant scholarship and the research presented here, that the leadership of American Indian women and Hispanas, whether in grassroots activism or in electoral politics, makes a difference to both the content of public policy and to the policymaking process.

> Women in [political leadership] affect not only who makes political decisions, but how and with what results. Women bring distinct experiences to the legislative debate and particular concerns to the policymaking agenda. Their presence makes a difference in what are considered political questions, and how those questions are resolved (Burrell 1994, 7).

As women's political involvement provides representation of previously marginalized groups' interests, it further serves to promote community participation in public life. To paraphrase Carole Pateman (1970), truly participatory democracy requires both the equal participation of citizens in decision-making as well as their equal sharing of power in determining the outcome of public policy. As Pateman notes,

> [t]he justification for a democratic system in the participatory theory of democracy rests primarily on the human results that accrue from the participatory process. One might characterize the participatory model as one where maximum input (participation) is required and where output includes not just policies (decisions) but also the development of the social and political capacities of each individual... (1970, 43).

My findings suggest that the 50 New Mexico leaders interviewed are devoted political actors who value democratic participation and work to improve social, environmental, and economic conditions in their communities. Unlike Olson's (1965) "rational actor," their participation is animated not by the prospect of selective incentives such as material benefits or the exercise of power, but by purposive/expressive incentives: the overarching goal of achieving economic equity and social justice for their communities.[14] The politics of these leaders are heavily influenced by their racial/ethnic- and gender identity as they seek to represent the interests of their ethnic communities, to advance the position of women in society, and to empower individual citizens to participate actively in public life. My findings reveal the prevalence of gender identity and racial/ethnic identity—not only as motives for women's political participation—but as factors that both inform and shape their public policy agendas within the arena of New Mexico's environmental politics.

CHAPTER 7
Conclusions

> When the composition of decision-making assemblies is seriously at odds with the gender and ethnic make-up of the society they represent, this means that certain voices are silenced or suppressed. If there were no substantial obstacles in the way of political participation, then those active in politics would be randomly distributed according to their ethnicity or gender; the fact that the distribution is far from random alerts us to these obstacles and the necessity for change (Longo 1997, 14).

THIS RESEARCH PROJECT EXAMINED THE POLITICS OF TWO GROUPS seldom studied within the discipline—American Indian and Hispanic women leaders—their political socialization, leadership trajectories, motives for engagement in public life, ideology, and racial/ethnic and gender identity, as well as their policy agendas and strategies for influencing policymaking. The context, environmental policymaking in New Mexico, provided an excellent opportunity in which to examine American Indian and Hispanic women's political participation. Not only do these racial/ethnic groups comprise nearly half of the state's population, but environmental issues are particularly salient for New Mexicans. Due to its arid climate, and a growing population that is increasingly concentrated in the two largest cities, the state must address problems related to water quality and quantity, and land use. Furthermore, we have a legacy of environmental exploitation and degradation. As a billboard near the Albuquerque airport proclaims, "Welcome to the only nuclear colony in the United States!"

The impact of the mining and nuclear industries on New Mexico's environment has no doubt contributed to the viability and saliency of the

Green party in the state and to the proliferation of grassroots environmental justice organizations here. Several of these latter groups have played instrumental roles in the development of the national environmental justice movement, which they continue to influence. Because these organizations are primarily headed by American Indian and Hispanic women, they provide excellent opportunities for women to acquire valuable leadership experience by developing their leadership skills and attributes, and by becoming acquainted with the public policymaking process (including gaining familiarity with the various policy forums, actors, and agencies). Moreover, as many women initially become politicized in grassroots or social movement politics before moving into leadership positions in electoral politics, we can follow their trajectories from grassroots activists to public office-holders.

Consequently, the comparative nature of this study facilitated examination of the variance and similarities between these groups while focusing on the effect that race/ethnicity and gender identity had on their experiences, ideology, and practice of politics. The findings contribute to our understanding of the political participation of American Indian and Hispanic women as leaders in grassroots organizations and in public office. Moreover, this study provides valuable insights to their politicization, motives for participation, ideology, political goals and policy agendas, and strategies for affecting change. I think it is helpful, at this point, to summarize the key findings relative to what we know from the existing literature. In this way, we can answer the questions: "What does this research add to our knowledge?" and "What difference does it make?"

SUMMARY OF THE RESEARCH FINDINGS

The indigenous and Hispanic New Mexico leaders were largely motivated to enter politics by a sense of civic obligation—the value they placed on "giving back" to the community through public service, by the opportunity to empower others, and by the desire to enhance the quality of life in their communities. This finding is consistent with what we currently know about much of the grassroots organizing activities undertaken by women of color (see for example Pardo 1998, 1995, and 1990; Kaplan 1997; Mack-Canty 1997; Sen 1995; Gilkes 1994; Di Chiro 1992; West and Blumberg 1990; Hamilton 1989; Kingsolver 1989). It also corresponds with the little we know about the politics of minority women in elected or appointed positions since, until recently, so few women of color have occupied these positions (see Glover 1998; Jacquez-Ortiz 1995; Cruz Takash 1993; Jackson 1993; Bonilla-Santiago 1991). In particular, my findings confirm those of other researchers studying American Indian and Hispanic women in public office (both tribal and non-tribal) as well as in non-governmental organizations (see for example Hoikkala 1995; Chiste 1994; Cruz Takash 1993; Hardy-Fanta 1993; Jaimes 1992; McCoy 1992; Miller 1992; Ford 1990; Koester 1988; Lynch 1986; Coyle et al. 1980).

My research, however, goes a step further by comparing two groups that are seldom studied together, American Indian women and Hispanas as political activists and public officials. As a result, we are able to gain some insights into the similarities and differences among these groups of women. For example, belief in their duty to serve their racial/ethnic community was especially pronounced among the indigenous women as well as among the public officials. Most of the leaders, regardless of their race/ethnicity or form of political leadership, sought to improve socioeconomic and environmental conditions and to empower community members by influencing public policy. These findings build on the literature of minority women's political activism, such as that cited above, while adding new knowledge about the policy preferences and strategies employed by women of color who seek to influence public decision processes.

The New Mexico leaders' motives for engaging in public service stemmed primarily from their childhood socialization into politics. Whether they were indigenous or Hispanic, the majority of the leaders were raised in families that were politically active. This was especially the case for public officials, more of whom reported having had role models or mentors and growing up in politically active families. That many of the American Indian and Hispanic leaders' families had long histories of involvement in electoral or grassroots politics may not seem surprising to individuals familiar with New Mexico's unique political history. Unfortunately, prevailing stereotypes of Indian women and Hispanas have traditionally painted them (and their racial/ethnic groups in general) as being apathetic, ignorant of, or completely detached from American political life (Liebowitz and Carroll 1996; Sierra and Sosa-Riddell 1994; Campos Carr 1988-89; Green 1980; Medicine 1978). My findings reveal that, at least for this group of New Mexico leaders, this is not at all the case. On the contrary, by achieving political leadership, many of these women were simply carrying on their family's tradition of participation and public service.

In addition to their families' influence and/or mentors' encouragement, the New Mexico leaders were further politicized by their involvement in service organizations, social movements, and/or electoral politics during their youth. As we might expect, more of the grassroots activists had been involved in social movement groups while more of the public officials had engaged in electoral politics. Overall, though, the vast majority of the leaders had volunteered for some sort of community service work, and still did, reflecting their deeply held belief in the value of giving back to the community.

While women of color in America often enter politics through community-based organizing activities, they may be less likely than these New Mexico women to have been raised in families with histories of leadership in electoral politics. In other words, the political trajectories of these indigenous and Hispanic leaders appear more similar to those of Euro American women than to other women of color around the country. This may be more a function of New Mexico's cultural make-up and political history, however, than an indication that these Indian and Hispanic women are more like Euro American women leaders than they are like other women of color in political leadership.

The fact that so many of the leaders interviewed came from political backgrounds in which several generations of family members had been active highlights the uniqueness of New Mexico politics. Unlike other states where minority populations such as American Indians and Hispanics have had to struggle for their share of political power, these groups have been recognized as legitimate players in New Mexico politics for some time (García 1994; Sando 1992; de la Garza and Vaughn 1985; Sierra 1984; Simmons 1979a and 1979b). This is not to say that these racial/ethnic groups share equally in political decision-making or influence. Certainly, Native Americans have had to fight for the right to participate freely in state and local politics (see chapter 1). However, insofar as Indians and Hispanics had their own well-established political and social institutions before colonization by the United States, their role as significant players in the public affairs of the state has been firmly entrenched.

Where racial/ethnic minorities and women have had to struggle to gain access and to develop influence in politics, they have been (at least initially) considered "outsiders" by the status quo. While, in many cases, Hispanas and native women continue to be perceived as political outsiders in New Mexico, their "outsider" status is largely a function of their sex and class rather than their race or ethnicity. The leaders in this study who held law degrees, who had been groomed for public office, and who carried the name of a well-established political family, no doubt enjoyed greater ease of access and were more readily accepted by the status quo than other leaders. Whereas these women were more likely to be perceived as and treated as political "insiders," native women and Hispanas from working class backgrounds who lacked the political connections and educational qualifications were more likely to enter grassroots politics. In combination, these factors almost certainly guaranteed their status as political "outsiders."

While there was some variation in social class among the leaders participating in this study, the differences were largely confined to the leaders' incomes and levels of education. More significantly, class differences were manifested in the ease of access and acceptance of the leaders into formal electoral politics. High social class tends to facilitate a political actor's achieving "insider" status. At the same time, while social class might affect how a leader is perceived by others in the political arena, class did not emerge as a characteristic that divided the 50 women studied here. On the contrary, there were many more important similarities than differences among the leaders such a their common goal of social justice, and their advocacy of racial equality and gender equity (see chapter 5). Moreover, there was substantial cooperation and interaction among the activists and public officials, which required crossing the boundaries of class, race/ethnicity, and political "insider/outsider" status (see chapter 6).

It appears from this study that, with respect to their political involvement, indigenous and Hispanic women leaders are highly motivated by concern for their racial/ethnic group and somewhat motivated by their desire to improve the status of women. This illustrates an important distinction from Euro American women leaders in that racial/ethnic identity and culture have tremendous salience for both native women and Hispanas whose concerns range from battling environmental racism, to preserving their native language, to maintaining tribal sovereignty. While this finding is not altogether new to the research on women of color, it is significant because it reiterates the necessity of having a racially and culturally diverse representation of women in politics. As one scholar notes,

> [e]thnic minorities may be more passionately and fairly represented by someone of their own group, and...their members in elected office provide valuable role models. When citizens can identify with their representatives they become less alienated and more involved in the political system (Burrell 1994, 6).

Seeing someone like oneself in leadership might certainly increase that individual's sense of political efficacy and perhaps even strengthen their support for the existing political system. If this was indeed the case, we might expect belief in the fairness of the system to ultimately lead to participation in politics. This is especially important if we consider that, until recently, women and racial/ethnic minorities were systematically excluded from participation in American politics.

Many of the New Mexico leaders described their political involvement as a deeply felt commitment to help others; politics was a means to make a difference for their communities. As a result, both the grassroots activists and public officials viewed their political work as less of a vocation and more as a personal responsibility or life purpose. The comments of Juana and Jackie, below, reveal the passion they felt for their work and are representative of the sentiments expressed by many of the women interviewed.

> This is what I do with my life. ...It's hard for me to separate myself from the organization. ...I'm a member of the community. What my

> work is, is who I am. [I'm] personally committed, and personally accountable to the community (Juana, activist).

> [My work has] fulfilled me because, for me, I don't separate out my job from my day-to-day life or family because my work is my family—it's the community I'm working for (Jackie, activist).

The leaders' goals centered on empowering members of their racial/ethnic group and women, and improving the quality of life for their racial/ethnic communities. These objectives were not only humanitarian but political as the intended results were to increase the participation and influence of these groups in public decision-making processes. The leaders' racial/ethnic identity and gender identity clearly played a prominent role in the formulation of their policy goals. In particular, racial/ethnic identity was more salient for indigenous women than for Hispanas and, while gender identity was less significant than racial/ethnic identity for both groups, it was slightly more important for Hispanas. As Hispanas are generally more integrated into the dominant Euro American society, this may mean that gender salience increases for women as their experience of racism or racial exclusion diminishes. What is especially interesting is that the New Mexico leaders expressed considerable gender consciousness.

As noted in chapter 5, gender consciousness influenced significantly their motives, ideology, and public policy agendas. The leaders identified with other women *as a group*, and many used their political activism to address women' issues by advocating for and/or empowering women. Indeed, several leaders had entered public life with this specific purpose in mind. None of them could be described as anti-feminists, rather, in various ways and to varying degrees, they all sought to improve the status of women in society. This finding is significant in that it reflects far greater support for a feminist agenda, at least among these women of color, than one might otherwise think based on the existing literature (see for example Junn 1997; Basu 1995; Pardo 1995 and 1990; Pesquera and Segura 1993; Trask 1993; Jaimes 1992; Bonilla-Santiago 1991; Johnson-Odim 1991; Ford 1990; Gunn Allen 1986; Aragón de Valdez 1980; Cotera 1980). It appears that, regardless of their race or ethnicity, women may have many more political interests in common that had previously been thought. This means that coalition-building across women's groups and organizing support for legislation affecting women may be less of a problem than anticipated.

The leaders' racial/ethnic identity and gender identity also shaped their political ideologies with respect to their identification with feminism and environmentalism. Based on the literature on women of color and feminism, which tends to depict women of color as highly critical of feminism (see above), it is surprising that a majority of the leaders self-identified as either "feminists" or "third world feminists." (See chapter 5). While there was little difference between the activists and officials with regard to their support of feminism, more of the native leaders identified with an alternative or "third world" feminism. Similarly, while the vast majority of activists and officials identified as either "environmentalists" or "third world environmentalists," greater numbers of American Indian leaders described themselves as "third world environmentalists." In general, these leaders rejected the mainstream Euro American movements in favor of alternative feminist and environmental ideologies that incorporated the cultural values, beliefs, and life situations of people of color into their political agendas. With one exception, I am unaware of any empirical research that has examined the political ideologies of native women and Hispanas with regard to their support for either feminism and/or environmentalism.[1]

These findings are significant for several reasons. First, they demonstrate greater support for both "mainstream" feminism and environmentalism by American Indian and Hispanic women than anticipated, given the literature. Second, my research reveals the ingenuity of the New Mexico leaders in adapting elements of these mainstream social movements to the needs of their own particular racial/ethnic communities. Furthermore, the alternative forms of feminism and environmentalism espoused by the leaders reflect their underlying concerns with social and environmental justice: preserving their culture, achieving racial and gender equity, increasing political participation, and maintaining or improving environmental quality.

Together, the motives, political goals, and ideology of the New Mexico leaders shaped their public policy agendas. While their agendas varied somewhat, depending on the women's race/ethnicity and form of leadership (whether they were activists or officials), they focused on three shared objectives: (1) to safeguard the rights of community members and empower them to participate in decision-making processes, (2) to secure services and resources to improve social and economic conditions for racial/ethnic minorities and for women, and (3) to promote the conservation of natural resources and the use of sustainable development practices. The New Mexico leaders did not view economic development and environmental quality as conflicting goals. They recognized the need for a strong economy yet were unwilling to compromise environmental health, so they advocated sustainable development and community participation in public decision-making processes.

While the public officials advocated more policies aimed at community development and the grassroots activists pursued greater numbers of environmental policies, the New Mexico leaders overwhelmingly supported policies to empower community members. Significantly, whether they were activists or officials, indigenous or Hispanic, the leaders' policy agendas reflected the goals of the environmental justice movement. As stated earlier, these goals incorporated such notions as gender equity, social, economic, and environmental justice, and increased participation in public decision-making. (See chapters 3, 5 and 6). These findings are significant because while they identify common themes, they confirm the results of a handful of studies that examine the political goals and/or public policy agendas of American Indian and Hispanic women (see for example Hoikkala 1995; Pardo 1995; Chiste 1994; Cruz Takash 1993; Hardy-Fanta 1993; McCoy 1992). When combined, these studies demonstrate that indigenous and Hispanic women leaders enter politics in order to improve the quality of life in their racial/ethnic communities by empowering community members and increasing participation in public decision-making processes.

Furthermore, my project reveals that there are far greater similarities among the policy preferences of native and Hispanic women leaders, as well as among grassroots activists and public officials, than might be expected from readings of classical political science literature. While scholars have traditionally assumed that significant differences exist between the public policy agendas of activists and office-holders, this does not appear to be the case where American Indian and Hispana leaders are concerned. In fact, there was greater convergence than divergence among the political agendas and strategies employed by the New Mexico leaders. Whether they were grassroots activists or public officials, these women similarly used a broad variety of strategies that incorporated both conventional and unconventional tactics for effecting policy change. (See chapter 6).

As many of the public officials cut their teeth in grassroots organizations before entering electoral politics, this finding is not surprising. Moreover, it illustrates important commonalities among the political goals, policy agendas, and strategies employed by indigenous and Hispanic women in New Mexico, and those of women of diverse cultures around the world (see for example Thomas-Slayter et al. 1996; Basu 1995; Chowdhury and Nelson 1994a; Bystydzienski 1992a and 1992b; West and Blumberg 1990). The implications of women sharing similar political goals and policy preferences across cultures and nationalities cannot be underestimated if we consider the potential for change of women's combined efforts. Support for women's international cooperation and unified action on issues such as environment, development, human rights, and public health is evident from the Beijing declaration and Platform for Action that grew out of women's participation in the United Nations Fourth World Conference on Women (see Women's Environment and Development Organization 1995).

IMPLICATIONS OF THE FINDINGS

Traditionally marginalized in the political science literature as passive, uninvolved, or simply absent from studies, it is only in recent years and because of women and racial/ethnic minority scholars that American Indian women and Hispanas have gained increasing visibility as legitimate and respected political actors in American politics. However, as this literature is developing, the current state of knowledge of the political leadership of indigenous and Hispanic women is limited. There is, as yet, much to learn. This project contributes in a modest way to the literature on political socialization, participation, political theory, gender politics, race/ethnic politics, and public policy by adding to our limited knowledge of the politics of American Indian and Hispanic women. Significantly, my findings provide empirical support for the importance of racial/ethnic minority women holding positions of political leadership.

Women do make a difference in how politics are played, in who gets to play, and in what concerns are raised. As one scholar notes,

> by underrepresenting women in government, we artificially narrow the government's views on policy and perhaps misrepresent the interests of women constituents. [Furthermore,] since women do not [share equally] positions of power in this country, women are led to believe that the political system is not open to them. ...Therefore, women could be more likely than men to question the quality of representation and the functioning of the U.S. system of government. ...Making the government more inclusive and increasing women's involvement would ensure that the talents and skills of all citizens were fully utilized in governing the nation (Kahn 1996, 138–139).

While the above quote refers to women in politics, it certainly applies to the representation of racial/ethnic minorities as well. For example, my findings show that American Indian and Hispanic women, while sharing substantial elements of their public policy agendas, do advocate for issues especially salient to their racial/ethnic group. Without their representation as public officials and grassroots activists, it is doubtful that their communities' particular interests (whether state, local, or tribal) would receive the attention that the New Mexico leaders gave them. Given the growth and potential influence of these racial/ethnic groups in national politics, the implications are significant for more responsive, and representative, federal and tribal policies.

If we truly believe in the value of a democratic system of government, we must actively seek to fill larger numbers of leadership positions with women of color—as is being done with increasing numbers of Euro American women in electoral politics. Based on the findings of this study, and on recent journalistic reports, American Indian and Hispanic women leaders in New Mexico are already meeting this challenge.

Table 26. Composition of the New Mexico State Legislature by Sex (1998)

	Senate	House	Total	%
Total	42	70	112	100
Females	10	20	30	27
Males	32	50	82	73

In 1998, New Mexico ranked 11th in the nation with women holding 27 percent of the seats in the state legislature, 10 in the Senate and 20 in the House, as illustrated by Table 26 (Glover 1998). While the figures remain disproportionate to the number of females in the state's population (51 percent)[2] as well as to the number of women registered to vote in New Mexico (54 percent)[3], the good news is that women's presence in electoral politics in the state is increasing. In fact, since 1989, the number of women in the New Mexico legislature has doubled.

Table 27. Sex of Candidates for Top Elected Offices in New Mexico (1998)

Office	Females	Males	Winner
Governor	0	8	male
Lieutenant Governor	2	2	male
Attorney General	2	4	female
1st Congressional District	2	5	female
2nd Congressional District	2	1	male
3rd Congressional District	4	6	male

Furthermore, as indicated by Table 27, greater numbers of women are running for elected office both in state and local government as well as in tribal politics (see Glover 1998; Linthicum 1998; Robertson 1995; Contreras and Shaw 1993). And, as more women run for office, they are breaking into positions of power traditionally reserved for men and transforming politics. As Burrell points out,

> of greatest effect on the political system...is the representation of women by women. The purposive activity of women within a legislative body on behalf of women's interests in the polity through the initiation of legislation to achieve equality is the single most significant aspect of women's presence as political leaders (1994, 151–152).

Therefore, based on my findings, I would alter the above quote to read, "Of greatest effect on the political system...is the representation of women and communities of color by women of color."

While women of color across the nation face significant barriers to achieving leadership and power in the political arena, native women and Hispanas in New Mexico may enjoy some unique advantages. These include a small state population comprised of relatively large numbers of racial/ethnic minorities; various influential grassroots organizations, which, as members of the environmental justice movement, are largely headed by women of color; an active network of women leaders; and numerous women already holding prominent elected and appointed positions in government. As American Indians, Hispanics, and other racial/ethnic minority groups become a growing force in American politics, leaders such as those featured in my study, increasingly serve as models for others seeking greater involvement in the political process. Further, studies comparing the effectiveness of grassroots activists and public officials in achieving policy change can provide direction for emerging groups that seek representation in public decision-making processes but who, as yet, may have limited political resources. Stated another way, reporting on the success of the strategies they employ and the achievements of such understudied groups as American Indian and Hispanic women leaders can offer other minority women, and men, the benefit of learning from their experiences.

Similarly, by examining the leadership trajectories of traditionally oppressed groups such as women of color, we may discover how to increase levels of political participation among other segments of the population. As population demographics in the United States change over time, our understanding of women of color in public office and in leadership positions in grassroots organizations should also change. The frameworks and research findings presented here can serve as an important point of departure for advanced studies thereby developing a richer understanding of these women as political actors within our system of democratic governance.

Likewise, by profiling role models such as the New Mexico leaders, scholars provide examples for citizens to emulate for their dedication to community service. Finally, as New Mexico politics provides a unique perspective from which to study the development and influence of the environmental justice movement on public policymaking, it serves as a model for examining the increasingly influential role of minorities and women in political leadership. Scholars interested in the future of American politics may wish to conduct research in such culturally and demographically diverse locations as New Mexico, California, Florida, Hawaii, and Texas because the populations of these states more accurately represent the "face" of America in the 21st century.

While the movement from political exclusion to political inclusion might remain a slow, arduous process, the results – more representative political institutions and political actors – are extremely desirable outcomes. This is especially salient for the women of color who serve the people of New Mexico and made this research endeavor viable. It is their political activism that makes our future more promising.

APPENDIX A

Roster of Project Participants

The following New Mexico leaders consented to their names appearing on this roster. Those not listed preferred to remain anonymous.

E. Shirley Baca	Theresa Baca	Phyllis Belindo
Ramona Carrillo	Eleanor Chavez	María C. Chavez
Judith M. Espinoza	Karen B. Foster	Jeanne Gauna
Stephanie Gonzales	LaDonna Harris	Dolores S. Herrera
Adele Hundley	B. Toby Chavez	Lela Kaskalla
A. Lehua Lopez	Loretta Naranjo Lopez	Linda Lovejoy
Sofía Martinez	Jobeth Mayes	Abby Moquino
Aleta Paisano-Suazo	Vickie S. Perea	Jill Peters
Martha Quintana	Darlene Smart-Herrera	Kathryn Tijerina
Gloria Tristani	Patsy Trujillo	Virginia Trujillo
María Varela	Linda M. Velarde	Carol Vigil
Rebecca Vigil-Girón	Verna Williamson-Teller	Esther Yazzie

APPENDIX B
Activist Interview Guide

Date:
Alias:
Leader's Name:
Length of Interview:
I want to guarantee your anonymity. I am using aliases rather than real names. No quote will be directly attributable to you.
Do I have your permission to tape record this interview?
Do I have your permission to use data from this interview for my dissertation and related research?
May I include your name in a roster of project participants?
Organization's name:
Organization's address:
Organization's telephone:
Title w/in organization:
Length of time in this position:
Length of time w/ organization:
Total time spent in public service:

1.0 What is your highest level of formal education?
1.1 Do you belong to a church or religious group? If so, which one(s)?
[If Native American, ask:]
Do you participate in your native religion or dances?
1.2 Are you married or currently involved in a committed live-in relationship?
1.3 What is your husband/partner's occupation?
1.4 Do you have any children under 18 years living with you? If so, how many?

1.5 Who has primary responsibility for housekeeping in your home?
1.6 Roughly, what percentage of housekeeping are you responsible for?
1.7 How old were you on your last birthday?
1.8 What was your total household income last year?
2.0 What language do you feel most comfortable speaking?
2.1 What language(s) did you grow up speaking?
2.2 What is your race or ethnic background?
2.3 What is your primary occupation?
2.4 Are you a primary income earner in your household?
2.5 How would you describe yourself?
3.0 Please describe your organization's purpose or mission.
3.1 What are your primary activities or duties?
3.2 What are your goals for the organization?
3.3 How does your organization do its business?
3.4 What strategies do you use to reach your organization's goals?
4.0 Why did you become involved in the organization?
4.1 How did you become involved?
4.2 A. How are decisions made?
B. Who makes the final decision?
4.3 How do you influence decision-making within your organization?
4.4 Are men and women equally involved in decision-making? Or is one group more influential?
5.0 What role do you play in the community? Please explain.
5.1 Have you had an impact within your community? How?
5.2 Have your efforts led to change? Explain.
5.3 Has your involvement in the community affected you? How?
Has your involvement affected your relationship with:
5.4 Your husband/partner? How?
5.5 Other family members? Who? How?
5.6 Friends? How?
5.7 Other members of the community? How?
5.8 What gets in the way of your community work?
5.9 What helps you with your community work?
6.0 When did you first become involved in issues outside of your home? Why?
6.1 Do you consider yourself political?
[If YES, ask 6.1 A, B. If NO, ask 6.1 X, Y]
A. What political activities are you involved in?
B. Of the people closest to you, who do you consider political? Why?
X. What is politics for you?
Y. Are any of your family active in community issues? Who? How?

6.2 Are you registered to vote?
6.3 Do you vote regularly?
6.4 Do you identify with a political party? If so, which one?
6.5 Are you a feminist? Please explain.
6.6 Are you an environmentalist? Explain.
6.7 Are you a political activist? Explain.
6.8 Are you a community organizer? Explain.
7.0 Is there anything else you'd like to tell me about your community involvement or yourself?
8.0 Last of all, which other Hispanic/Chicana and Indian women—who are involved in environmental issues—would you suggest I contact to interview? I need their organizations and phone numbers as well as their names.

Thanks so much for taking the time to speak with me.

APPENDIX C

Public Official Interview Guide

Date:
Alias:
Leader's Name:
Length of Interview:
I want to guarantee your anonymity. I am using aliases rather than real names. No quote will be directly attributable to you.
Do I have your permission to tape record this interview?
Do I have your permission to use data from this interview for my dissertation and related research?
May I include your name in a roster of project participants?
Name of Agency/Commission/Elected body:
Official address:
Official telephone:
Official Title:
Elected/Appointed Position:
Dates of term/service in present position:
Total time spent in public service:

1.0 What is your highest level of formal education?
1.1 Do you belong to a church or religious group? If so, which one(s)?
[If Native American, ask:]
Do you participate in your native religion or dances?
1.2 Are you married or currently involved in a committed live-in relationship?
1.3 What is your husband/partner's occupation?
1.4 Do you have any children under 18 years living with you? If so, how many?

1.5 Who has primary responsibility for housekeeping in your home?
1.6 Roughly, what percentage of housekeeping are you responsible for?
1.7 How old were you on your last birthday?
1.8 What was your total household income last year?
2.0 What language do you feel most comfortable speaking?
2.1 What language(s) did you grow up speaking?
2.2 What is your race or ethnic background?
2.3 What is your primary occupation?
2.4 Are you a primary income earner in your household?
2.5 How would you describe yourself?
3.0 What are your primary activities or duties in the (official body)?
3.1 How do you influence decision-making in the (official body)?
3.2 Are men and women equally involved in decision-making? Or is one group more influential?
3.3 Why did you become involved in politics?
3.4 How did you become involved in politics?
3.5 What is the (official body's) mission/purpose?
3.6 How does the (official body) do its business?
4.0 Which policy issues (or programs) are particularly important to you? Why?
4.1 What do you want to accomplish regarding these issues?
4.2 What strategies do you use to influence policy?
4.3 What role do you think community activists play in the policy process?
4.4 Do you think activists help or hinder the political process? Please explain.
5.0 Have you had an impact in the community since you became involved in politics? Explain.
5.1 Have your efforts led to change? Explain.
5.2 Has your political involvement affected you? How?
Has your involvement affected your relationship with:
5.3 Your husband/partner? How?
5.4 Other family members? Who? How?
5.5 Friends? How?
5.6 Other members of the community? How?
5.7 What gets in the way of your political work?
5.8 What helps you with your political work?
6.0 When did you first become involved in issues outside of your home? Why?
6.1 What is politics for you?
6.2 Of the people closest to you, who do you consider political? Why?
6.3 Are you a feminist? Please explain.
6.4 Are you an environmentalist? Explain.

6.5 What is the difference between a community activist and a politician?
6.6 Political party membership
7.0 Is there anything else you'd like to tell me about your political involvement or yourself?
8.0 Last of all, which other Hispanic/Chicana and Indian women – who are involved in environmental issues – would you suggest I contact to interview? I need their organizations and phone numbers as well as their names.

Thanks so much for taking the time to speak with me.

APPENDIX D
Human Subjects Review Consent Form

University of New Mexico
College of Arts & Sciences
Human Subjects Institutional Review Board

CONSENT FORM TO PARTICIPATE
IN A DOCTORAL DISSERTATION RESEARCH PROJECT

This project has been explained to me by Diane Prindeville, a doctoral student in political science at the University of New Mexico, conducting research on women in politics. I am participating voluntarily. I can withdraw my consent to participate at any time without penalty. I have been guaranteed anonymity and understand that no quote will be directly attributable to me. Any questions I have can be answered by Diane Prindeville at (505) 268-3452 or (505) 277-5104.

______________________________ ________________

Participant's Signature Date

I also agree to my name being listed in a roster of project participants.

____

Participant's Initials

Notes

NOTES TO CHAPTER 1

[1] The terms "Hispanic" and "Hispanas" are used throughout to refer to Mexican American women and to women of Spanish heritage residing in New Mexico. Mestizas, women of mixed Spanish and Indian ancestry, are included in this group. Women enrolled as members of a Pueblo, tribe, or Indian nation are referred to here as "American Indian," "native," "indigenous," or "Indian" - the vernacular in New Mexico.

[2] The phrase "of color" is used throughout to refer to members of minority racial and ethnic groups in the United States who are neither primarily Anglo- nor Euro American.

[3] In 1990, Hispanics comprised over 38 percent of the state's population. However, numerous sources within New Mexico insist that the Hispanic population was grossly undercounted in that census. Also, the state has the second largest number of American Indians as a percent of total population (9 percent) (U.S. Department of Commerce 1990a, Table 28).

[4] Etienne and Leacock differentiate between horticultural "gardening with hand tools" and agricultural "elaborated cultivation, often with ploughs, on lands prepared by terracing, drainage, or irrigation" societies (1980, 8).

[5] Although descent is traced through females in matrilineal societies, one must be cautious not to automatically ascribe power to women as, even in this system, position and power are often controlled by men. Matrilineality is not necessarily associated with matrilocality, in which married couples reside with or nearby the wife's mother (Seymour-Smith 1986, 185).

[6] This has been documented for women in other tribes, as well. See, for example, Maltz and Archambault (1995), Ford (1990), Albers (1983), Niethammer

(1977). For a fascinating and detailed discussion of the role of American Indian women in native societies prior to European contact, see Albers (1989).

[7] For more of the history of native-Spanish relations during colonization, see Gutierrez (1991), Brugge (1983), Lange (1979), Ortiz (1979), and Simmons (1979a).

[8] Early Spanish occupation was marked by conflict and native resistance, culminating in the Pueblo Revolt of 1680. In the years that followed, however, the Spanish both regained and maintained political, social, and economic control of the region. For a chilling and well documented account of the atrocities committed against native peoples on behalf of Spain and the Roman Catholic Church, see Gutierrez (1991).

[9] For a comprehensive explanation of this process, and of the anthropological terms used here, see Etienne and Leacock (1980). In addition, it is important to acknowledge that colonization did not impact all native women in the same way. On the contrary, whereas women's status declined substantially among some Indian tribes, in other regions women's status may have risen depending on local economic conditions. For an excellent comparative analysis of the experiences of native women under colonization, see Albers (1989).

[10] The feminine played at least an important symbolic role in the performance of both native and Christian religious and cultural practices. For example, both Indian peoples and *pobladores* affirmed women's contributions to their spiritual life through celebrations of the feminine (or mother figure) in religious ceremonies and festivals. Arguably, *la Virgen de Guadalupe* [the Virgin of Guadalupe] and lesser female saints paralleled the significance of the Corn Mother and other female entities in native belief systems (Rebolledo et al.1992; Gutierrez 1991; Gunn Allen 1986).

[11] Numerous sources indicate that the great majority of the "Spanish" settlers were actually mestizos - racially mixed Mexican born descendants of Spanish *conquistadores* (see for example Gutierrez 1991; Simmons 1979a; Gonzalez 1969). It is therefore more accurate to refer to these communities as Spanish/Mexican.

[12] By "agropastoral" I mean an economy based on subsistence farming and on raising small herds of sheep, cattle, and/or horses for both family use and trade.

[13] These particular activities were not the exclusive domain of either women or men, however. Indeed, in some tribes weaving was a "female" activity while in others it was "male." On the other hand, work such as harvesting crops or maintaining dwellings might involve the community as a whole.

[14] This was generally the case in other areas where native peoples similarly experienced the social and economic effects of colonization (Chiste 1994; Devens 1992; Jaimes 1992; Bonvillain 1989; Etienne and Leacock 1980).

[15] While they endured oppressive social conditions, women in New Spain enjoyed particular legal rights that women in colonial New England did not. Under

the Spanish system of law, free women were entitled to own and inherit property in their own names, independently from their husbands. They had the right to make wills, and they could litigate in court.

[16] While this was certainly the case for the Pueblos of the Rio Grande valley, it was not so for the Zuni and the Hopi who were somewhat more insulated from the effects of Spanish Colonization. "Genízaras/os" were generally the offspring of Apaches, Navajos, or Comanches who had been captured in battle and enslaved. While these captives were marginalized by both the Pueblos and the Spanish settlements, where their labor was essential to the maintenance of the colonial economy and social class structure, their children were considered free. For an in depth discussion of *genízaros*, see Chavez (1979). Also see Lange (1979).

[17] The issue of racial and ethnic identity is complicated by historical prejudices which even today lead many individuals to claim their Spanish heritage while denying their native roots (Stiffarm 1992). Some scholars argue that few contemporary New Mexicans can claim "pure" blood - whether Indian or Spanish because the majority of the non-Anglo population is in fact mestizo/a (Gutierrez 1991; Gonzalez 1969). Also see Simmons (1979a).

[18] It is important to note that unlike the Navajos, for example, the Pueblos were not displaced from their land.

[19] One source notes that "[i]n the Federal Indian schools, physical abuses, beatings, imprisonment, forced labor, and malnourishment were chronic conditions" (Willard 1984, 72). Also see Kelley (1999).

[20] American Indian women were excluded as native peoples were barred from voting in state and local elections at this time.

[21] While she does not discuss New Mexico specifically, Cotera (1980) provides a wonderfully detailed history of the political activism of Chicanas in the Southwest from 1848 into the 1970s.

[22] For a folksy yet comprehensive discussion of New Mexico politics from the turn of the century through the 1960s, see Holmes (1967). For a more recent and critical analysis of the state of New Mexico politics, see García (1994).

[23] The General Allotment Act of 1887, also called the "Dawes Act," was one of many federal policies of forcible assimilation, designed, in this case, to break up Indian land holdings and tribal unity (see McClain and Stewart 1995; Churchill 1992; Churchill and Morris 1992; Robbins 1992). For a more detailed discussion of Pueblo-U.S. relations, see Simmons (1979b). For more on the impact of the Indian Reorganization Act on particular tribes, see Rusco and Rusco (1986), Iverson (1983), Shepardson (1983), and Simmons (1979b).

[24] (25 U.S.C. 461 et seq. 48 Stat. 984 Pub. L. 73-383 Pub. L. 103-263). This information was culled from the following websites:

http://www.doi.gov/bia/stats930.html

http://www.fw.umn.edu/indigenous/gloss.html

http://www.library.arizona.edu/branches/spc/udall/hidden.html
http://www.msnbc.com/onair/nbc/nightlynews/fleecing/newdeal.asp

[25] Pueblo forms of government were also influenced by Spanish colonial structures. For a more extensive discussion of the various Pueblos' governmental systems, see Sando (1992) and Simmons (1979a and 1979b).

[26] Ironically, because of the blurring of religion and politics, several Pueblos continue to exclude women (and lay men) from participation in tribal politics. As Albers notes, in many tribes, religious ceremonies were "situations where gender boundaries were inviolable" (1989, 138). In the case of the Pueblo tribes, traditional religious practices have provided the justification for maintaining exclusionary political systems up to the present. For more regarding Pueblo systems of government, see Eggan (1983 and 1979), Iverson (1983), and Simmons (1979a and 1979b).

[27] Some tribes have constitutions while others have tribal codes (similar to a system of municipal ordinances), which carry the force of law. It is often difficult to obtain accurate information about Pueblo systems of governance as there is a certain amount of secrecy surrounding their tribal administrations. For example, I found that project participants were forthcoming and cordial except when it came to issues relating to specific tribal political practices.

[28] Albers explains that this has been due, in large part, to the ability of the Navajo to "maintain some of the essential features of their indigenous economy and social organization," as they are traditionally a matrilineal society in which women have enjoyed significant property rights and have held a principal role in production (1989, 146). For a comparison of women's gender roles and relative power in Pueblo and Navajo society, see Jacobs (1995) and Shepardson (1995), respectively.

[29] Chapters within the Navajo Nation are political jusrisdictions roughly equivalent to rural counties.

[30] As one researcher concludes in her study of women tribal leaders,

> it is the Anglo culture which draws such sharp distinctions between the proper roles of men and women in public life, and if tribal women do exhibit different political behaviors than men or do define their political roles differently than do men, it may be more as a result of the imposition of Anglo culture and Anglo government structure on the tribes than the tribal cultures themselves (McCoy 1992, 67).

[31] "The personal is political" became a rallying cry for the women's liberation movement. As women around the country gained awareness of their shared social, economic, and political situations, they began seeing themselves as a class of people who, because of their sex, were subject to disparate and unjust treatment by individuals and institutions. Concerns that had formerly been considered private,

such as spouse abuse, sexual harassment, childcare, and rape, were now understood as being legitimate issues that deserved political consideration (see Sapiro 1981).

[32] The historic women's movement, which concerned itself with numerous economic and legal initiatives in addition to obtaining suffrage for women, is commonly known as the "first wave" of feminism. The rebirth of feminism in the United States in the 1960s and seventies is referred to as the "second wave."

[33] These criticisms are addressed in greater detail in chapter 5.

[34] I am referring to the economics of providing for the material well-being of one's family with the finite and often scarce resources that many women have to work with.

[35] For example, while stereotypes depict mothers as knowledgeable of children's health issues simply because they *are* mothers, their knowledge of environmental toxins may be discounted because they are perceived as *only* mothers and therefore lacking in expertise outside of their domestic role. See especially Bellows (1996).

[36] The United Nations' 1995 *Report of the Fourth World Conference on Women* provides a dramatic example of their influence. For a synopsis, see "Turn the Words into Action! Highlights from the Beijing Declaration and Platform for Action," November 30, 1995, published by the Women's Environment and Development Organization (WEDO), New York. Other women's international policymaking organizations include the International Policy Action Committee, the International Women's Health Coalition, and the Committee on Women, Population, and the Environment. Respect for women's rights is one common goal of these groups (see Rodriguez-Trias 1994). Also see the work by additional authors in West and Blumberg's (1990) anthology.

[37] This is supported by the data in my own study as discussed in Chapter 6.

[38] To illustrate: In her study of Latinas in elected office, Cruz Takash (1993) found that many of the leaders she interviewed had considerable political experience as former grassroots activists, yet they continued to struggle with obstacles based on sex, race/ethnicity, and class, as well as internal cultural impediments (see also Bonilla-Santiago 1991; Aragón de Valdez 1980). Other research demonstrates that American Indian women face the same challenges (Linthicum 1996; Jaimes 1992; Gunn Allen 1986).

[39] The so-called "gender gap" was first observed, and the term was coined, during the Reagan administration when pollsters noticed significant differences between the support of female and male voters for President Reagan and his policies (men favored them far more than did women). The gender gap was evident again during the 1992 election when Bill Clinton won the presidency and significant numbers of women were elected to Congress.

[40] It appears, however, that while women are experiencing greater success at winning elected office and influencing public policy, they have a long way to go

toward leveling the political playing field, achieving parity with men, and gaining entré to the highest echelons of political leadership (Fox 1997; Dodson et al. 1995; Thomas 1994; Carpini and Fuchs 1993; Deutchman 1992). As Witt, Paget, and Matthews wryly observe, "[t]he trick is to bring more women into politics, and to bring more of the values traditionally associated with them into the political arena, without establishing a permanent link between the two" (1994, 283).

[41] S. J. Carroll's original study was conducted using data from women's political campaigns run in the 1976 elections. Although this raises questions about the applicability of her conclusions two decades later, research published since then has supported her most important findings (see for example Havens and Healy 1991; Saint-Germaine 1989; Darcy et al. 1987; Flammang 1984a; Rule 1981). The second edition of Carroll's book, which I cite here, includes analyses from the 1992 election cycle, popularly termed "The Year of the Woman."

[42] While the authors do not specifically mention acculturation, they imply that increased levels of political participation among Mexican American women are due to their successful acculturation.

NOTES TO CHAPTER 2

[1] According to the 1990 census, the total number of American Indians, Eskimos, and Aleuts was 128,068 or 8 percent of the state's population. Combined with all persons of Hispanic origin (579,224 or 38 percent), these groups account for 707,292 persons or approximately 47 percent of New Mexico's population (United States Department of Commerce, Bureau of Census 1990a).

[2] Recent work, such as that of Fox (1997), Dodson et al. (1995), Burrell (1994), Thomas (1994), Witt, Paget, and Matthews (1994), have all included relatively large samples of American women leaders but do not address issues of race or ethnicity in their research. Similarly, Fowlkes (1992) included both grassroots activists and elected and appointed officials in her study but limited her sample to white American women.

[3] The particular policy objectives of the leaders will be discussed in detail in chapter 6.

[4] Scholars such as Hardy-Fanta (1993), Fowlkes (1992), and Ford (1990) have included both grassroots activists and public officials in their examinations of women's political leadership. I am unaware, however, of any studies that compare or contrast the politics of American Indian women with that of Hispanas.

[5] Chapter 3 provides a discussion of both my findings and the literature, which support this claim.

[6] United States Department of Commerce, Bureau of Census (1990a).

[7] I say "safely assume" because throughout history, as in the present time, women have generally been excluded, both through legal means and by custom, from holding positions of political leadership. Chapter 3 provides a more detailed discussion of the history of women in New Mexico politics.

[8] All participants were guaranteed anonymity if they so chose it but many allowed their names to be listed in a roster at the front of this document. I have used pseudonyms and edited-out any identifying material from their quotes to preclude attribution to individual leaders and to respect their requests for privacy.

[9] Chapter 3 follows with an in-depth discussion of the environmental justice movement in New Mexico, its history and goals.

[10] According to Krauss, politicization is a process by which

> individuals develop a framework of meanings and beliefs that challenge dominant ideologies and empower political action. ...The construction of new interpretive frameworks is a subjective process through which people attribute their discontent to structural, cultural, or systemic causes, rather than to personal failings or individual deviance (1998, 131).

[11] I agree with Holstein and Gubrium that: "Because socially constructed meaning is unavoidably collaborative, it is virtually impossible to free any interaction from those factors that could be construed as contaminants. All participants in an interview are inevitably implicated in making meaning" (1995, 18).

[12] For an in-depth discussion of feminist activist scholarship, see Naples (1998a). (Also see Kaplan 1997; Liebowitz and Carroll 1996; Benmayor 1991; Berger Gluck 1991; Fonow and Cook 1991; Patai 1991).

NOTES TO CHAPTER 3

[1] Responding in 1947 to the displacement of Native peoples by the federal government, activist Amy Hallingstad ironically noted:

> The men from Washington have set aside many millions of acres on which wolves and bears may not be disturbed and nobody objects to that. Perhaps if we were wolves or bears we could have as much protection. But we are only human beings. There are no closed seasons when it comes to skinning Alaska Natives (Koester 1988, 38).

Hollingstad's comments remain relevant as they reflect the views of many indigenous peoples towards both the federal government and mainstream environmentalist groups.

[2] This woman, whose activism on behalf of old growth forests has resulted in her receiving numerous death threats, asked not to be identified.

[3] In a study that examined alternative forms of compensation to community residents in five Massachusetts towns faced with the hypothetical placement of hazardous waste treatment facilities, researchers found overwhelming support for risk mitigation proposals by residents over a number of economic incentives offered to mitigate public opposition to the sites (Portney 1985. See also Bullard 1990).

[4] For a more detailed discussion of SNEEJ and similar organizations working in Mexican American neighborhoods, see Marquez (1994).

[5] Also see Kaplan (1997), Mack-Canty (1997), Rocheleau et al. (1996), and Pardo (1995).

NOTES TO CHAPTER 4

[1] Cruz Takash (1993) studied 76 Hispanas in public office in California, 94 percent of whom were elected. While my sample included 11 Hispanic elected and appointed leaders in New Mexico, it is interesting to compare the profiles of the leaders interviewed. Like the New Mexico leaders, the majority of the California leaders were 40 years or older. Not surprisingly, the household incomes of the California leaders were somewhat higher with the majority (62 percent) reporting an income between $50,000 and $100,000. In contrast, 73 percent of the New Mexico leaders had a household income of $35,000-$75,000. This may be explained, in part, by the fact that 78 percent of the California leaders were married and the majority (52 percent) had grown children. Fewer of the New Mexico leaders, on the other hand, had live-in partners (55 percent) and 36 percent had minor children. Furthermore, California, one of the wealthier states, contrasts sharply with New Mexico, one of the poorest in the union, with regard to household income. Despite their lower levels of income, however, 36 percent of the New Mexico leaders had some college or a Bachelors degree and 45 percent had a juris doctor or PhD degree, while 22 percent of the California leaders reported having less than an Associate of Arts degree and 21 percent had a Masters degree. Finally, while 65 percent of the New Mexico leaders remain bilingual in Spanish and English, 53 percent of the California leaders reported that English is their primary language (Cruz Takash 1993, 342–343).

[2] McCoy (1992) studied 19 American Indian women leaders in tribal government. My sample included 25 Native American leaders, of whom only five were tribal officials; the remainder were either public officials in national, state, or local government, or grassroots activists. However, it is still instructive to compare these groups. In McCoy's study, the average leader was 52 years old with between 0 and two children, and the majority (63 percent) were married. Furthermore, 37 percent had a college degree, 26 percent had two years or less of college, and 37 percent had a high school diploma (1992, 60). In contrast, of the 13 Indian officials I interviewed, the average age was 46, 77 percent had a live-in mate, 46 percent had minor dependents, 31 percent had two years or less of college, 38 percent had a bachelors degree, and 31 percent had either a Masters degree, a juris doctor degree, or a PhD. Overall, comparing McCoy's sample of tribal leaders with the New Mexico Indian officials, the latter group was generally younger, better educated, and proportionally more were partnered.

[3] For example, their mate or other family member(s) contributed to the household income, or the leaders themselves held a paying job in addition to their political work.

[4] Nevertheless, as S. J. Carroll notes in a more recent study, indicators of the opportunity structures available to women interested in entering politics remain "critical in explaining the lack of greater numbers of women among political elites" (1994, 5).

[5] In a study comparing female and male public officials, S. J. Carroll found that

> women appointees...were far more likely to be unmarried and childless than their male counterparts suggest[ing] that the women far more often than the men had chosen to delay marriage and childrearing, or to bypass marriage and childrearing completely, in order to pursue careers. Most of the men had pursued both family and career simultaneously, suggesting that career decisions were made without a perceived need to choose between career and marriage or children (1989a, 61).

These findings are similar to those concerning appointed officials in state government (see Havens and Healy 1991).

[6] For a breakdown of the specific racial/ethnic composition of this group of leaders and how they compare with women in the New Mexico public, see Table 1 and the discussion of the sample in chapter 2.

[7] In her study of Euro American women, Fowlkes (1992) also found that childhood socialization, consciousness-raising experiences, mentors, and what she termed "countersocialization" had variable impacts on the degree of involvement and the ideological orientation of white women's political activism. Fowlkes defined countersocialization in terms of its content on two dimensions - roles and values - and posited that

> members of groups defined as dominant or subdominant on the basis of sex and race may nevertheless strive to develop their own modes of action, or subdominants may strive to occupy roles reserved by tradition or law for dominants, because the individuals have developed self-concepts and/or values that run counter to dominant sociopolitical norms (1992, 15–166).

[8] Interestingly, research comparing Afro American- and Euro American women found that the political ambition of the latter group was tied to "nontraditional sex role beliefs" - Fowlkes' (1992) notion of "countersocialization" - while the former group's ambition was tied to "current activities and the existence of parental interest in politics" (Perkins 1986, 27). Both groups of women, however, had been countersocialized growing up. The author concluded that political socialization by racial/ethnic minority families may be more important to their daughters' political ambition and participation in adulthood than it is for white women who have greater opportunities in the dominant society.

[9] It is important to note that the New Mexican political experience may be atypical. A study of Chicano political socialization found that New Mexican respondents

> ...were significantly more likely [than Chicanos in other Southwestern states]...to have fathers whose occupations reflected higher socioeconomic status... [Furthermore,] New Mexican respondents [were] much more likely to have relatives who are active in politics...and to have become politically active as a result of family traditions... (de la Garza and Vaughan 1985, 250).

This is further supported by Cruz Takash who found that only 14 percent of the Hispana leaders whom she interviewed in California had been raised in politically active families (1993, 346).

[10] I use the term "constituent" loosely here to refer to citizens represented by an activist's organization or a public official's office.

[11] A study of Scottish women's political socialization found that their involvement in sex-segregated groups had a significant effect on raising women's political consciousness, regardless of the nature of the organization (Chapman 1987). When women participated in exclusively female associations, they gained a "perspective which makes them conscious of women as a political category and relates women's interests to the political processes in which they are engaged" (330). Interestingly, several of the New Mexico leaders reported participating in all-female organizations of various types.

[12] Paehlke and Rosenau (1993) note that despite some clashes of opinion, environmental and labor union organizations have actually enjoyed mutually supportive relations. The authors point out that there

> has been strong agreement on the need to expand the solar industry, on occupational health protection, on the need for and the means of achieving full employment, on pollution abatement, on energy conservation, and on a wide range of other pro-environmental policies (1993, 681).

[13] The difference in the involvement of native leaders and Hispanas in social movement organizations may have to do with the existence of larger numbers of groups that addressed Mexican American issues. In contrast, only AIM had national recognition and represented specifically native people's concerns, and even so, it did not receive great support from members of Southwestern tribes.

[14] In an early study, Orum identified four factors influencing one's decision to participate in a "political protest movement:" (1) the individual's socialization into a political belief system, (2) their relative deprivation or subjective dissatisfaction with the existing social, economic, and/or political structures, (3) their sense of political efficacy, and (4) an unstructured work routine allowing for free time (Orum 1974, 199-201). While his research did not specifically examine the impact of race, ethnicity, or gender, Orum's model provides a useful, if basic, framework for examining an individual's motives for participation in grassroots political organizations.

[15] Fowlkes (1992) provides an interesting analysis of Euro American women's understanding of "the political" and how these conceptualizations influence their choice of political engagement.

[16] American Indian support for the Republican party largely originated with the Nixon administration's recognition of native concerns during the 1970s and their involvement in returning Blue Lake (an important spiritual site in northern New Mexico) to Taos Pueblo.

[17] The term "outside" was sometimes used by native women to refer to life off-reservation.

[18] A 1984 study of women's political candidacies in New Mexico revealed increasing numbers of women running for nontraditional offices (Clark and Clark 1984). More recently, articles in the local newspaper indicate that not only are more women running for office, they are winning an increasing number of elected and appointed positions in state, local, and tribal government (Glover 1998; Linthicum 1998) with the help and financial support of other women and of women-run political action committees (Paul 1995).

NOTES TO CHAPTER 5

[1] For a thoughtful and more detailed discussion of the fine points of racial/ethnic identity in the Southwest, see Lamphere, Zavella, and Gonzales (1993).

[2] While only 20 of the 50 leaders used racial/ethnic terms when describing themselves to me, I think this rather low number is misleading. It may be due to the fact that the women participating in my study knew that they had been selected *because* of their racial/ethnic identity. I began each of my interviews with a brief explanation of my project, the focus of which, I said, was the leadership of American Indian and Hispanic women in the state's environmental politics. Furthermore, I identified my own interest in the project as a Latina woman. Therefore, it may have seemed redundant to some of these leaders to describe themselves to me by their racial/ethnic identity.

[3] I am borrowing Sue Tolleson Rinehart's (1992) concept of "gender consciousness" and adapting it. Her notion of gender consciousness is discussed in greater detail below.

[4] For more information on the subject, see Prindeville, Gonzales, and Sierra (1992). Our report, *The State of Ethnic and Race Relations in Albuquerque, New Mexico*, was funded by and prepared for the Levi Strauss foundation's Project Change, a multi-city program to improve community race relations.

[5] Cruz Takash (1993) similarly documents the incidence of racism and sexism experienced by Hispanas in elected office.

[6] In New Mexico, the offices of County Clerk and Secretary of State are traditionally held by women.

[7] The varying status of native women may be due to their tribes' diverse experiences of and responses to colonization. Albers (1989) explains how the status of native women in different regions was affected by the political economy. The result-

ing changes in women's and men's roles and their control of resources led to Plains women, for example, losing status while women from egalitarian tribes in the Great Basin retained theirs.

[8] In her study of women tribal leaders, McCoy found that four of the 19 leaders interviewed believed that their tribe's cultural traditions discouraged both women *and* men from participating in politics because "competitive values are not instilled in either men or women in some tribes, making it difficult for either to compete for offices in a tribal government structure imposed by Anglos" (1992, 66); *not* because tribal traditions exclude women, in particular, from involvement in tribal politics, as do some of the Pueblo nations.

[9] S. J. Carroll raises an important point about the causal relationship between consciousness-raising and activism noting that the direction of the relationship may go either way. "While feminist attitudes may motivate political activism, they also may result from participation. Socially interactive participation may itself be an agent of socialization and perform a consciousness-raising function" (1989b, 329).

[10] As Pesquera and Segura note, "The Chicano Movement exalted marriage and reproduction as integral to the politics of cultural reaffirmation. A fundamental feminist position, on the other hand, indicted marriage and reproduction within the traditional patriarchal family as a primary source of all women's subordination" (1993, 99).

[11] Cruz Takash found that while a minority of Hispana officials (33 percent) in her study described themselves as feminists, the majority (51 percent) agreed with the statement "I am somewhat ambivalent about feminism, do not call myself a feminist but support many of the issues feminists advocate" (1993, 351). Fewer than 12 percent of her sample did not support feminist policy agendas. Most significantly, the vast majority (84 percent) of the Hispana leaders interviewed "claim[ed] to support feminist goals" (1993, 350).

[12] While gender consciousness manifested itself in the New Mexico leaders' activism as advocacy for women's rights, it is important to acknowledge that women's awareness of their gender interests does not necessarily result in feminist activism as evidenced by the involvement of women in highly conservative right wing groups in the United States and abroad. As one scholar states,

> consciousness itself is neither uniform nor monolithic. Often women's experience of their identities as gendered subjects may not assume a feminist form. ...[T]he growing consciousness that identities are multiple rather than singular has often led women to emphasize race, ethnicity, and sexual orientation as much as gender (Basu 1995, 17. See also Orleck 1997; Fowlkes 1992).

[13] This finding is not limited to women of color as white working class women have also rejected the label "feminist" and shown hostility towards the women's movement while supporting women's rights to equal pay, for example. Cable

explains this as illustrating "the salience of class over gender issues" (1992, 46. See also Orleck 1997; Fowlkes 1992; West and Blumberg 1990; Dabrowski 1983).

[14] In a 1993 study, Cruz Takash similarly asked Chicana and Latina elected officials in California to describe themselves with regard to feminism. Of the 76 leaders surveyed, nearly 33 percent self-identified as feminists, 51 percent supported "many of the issues feminists advocate" although they themselves were ambivalent about feminism, and less than 12 percent rejected feminism. The leaders were not specifically asked whether they supported an alternative form of feminism. While 54 percent of the New Mexico leaders embraced feminism, 4 percent were ambivalent, and 25 percent rejected feminism. (See Table 17). However, 17 percent identified as third world or Chicana feminists and the 25 percent who rejected feminism nevertheless supported issues that feminists advocate. In contrast with the California leaders, a greater percentage of the Hispana leaders from New Mexico either identified with some form of feminism (71 percent) or supported feminist principles (29 percent), despite their rejection of feminism as a movement.

[15] For an overview of the varying theories and forms of feminism, see Tong (1989).

[16] One activist credited the local chapter of a national environmental group with politicizing her. "The Sierra Club...got me motivated to organize in [my community] to prevent the cementing of the *acequias* [irrigation canals]. It was a threat to our way of life, our history. It would destroy the cottonwoods and the willows. It was a moving force for me" (Lucia, activist).

[17] It is important to distinguish between "preservation" and "sustainable use" or "conservation," as these terms are often confused. The preservation movement grew into the mainstream environmental movement, which stresses the value of maintaining wilderness areas and biodiversity for their own sake, whether or not their existence directly benefits human beings. In contrast, "sustainable use" or "conservation" (I use these terms interchangeably) involves the development and "wise use" of natural resources and places human needs above those of non-human life. Caulfield, who has written extensively on the history of the conservation and preservation movements describes them in the following ways.

> The conservation movement focused its concerns upon natural resources, that is, those things in the environment which are of material, economic concern to man (sic). Constraint in their use should be only that degree necessary to assure their long-term availability for material use (1989, 49).

Caulfield goes on to describe conservation as the "multiple use sustained yield of renewable resources" (1989, 50). The principles underlying the movement include a belief in conservation as "the greatest good, for the greatest number, for the longest time" and the notion that "the federal public lands belong to all the people" (1989, 20–21). Among the public policy objectives of the conservation movement were "public and private forest management [and] comprehensive multiple-purpose river basin development" (1989, 23). In contrast, preservation has, "as an object of its ideological concern, the whole geosphere and biosphere including man (sic) himself. The continued long-term integrity of this whole object of concern is its central value" (1989, 49). Furthermore, the preservation movement

> strongly insists that multiple-use means respect for wildlife and recreation uses, ...[they] favor greater governmental control of private land and the purchase of private land, where necessary, to achieve environmental values (1989, 50), [and they] strongly favor solutions to pollution problems even if higher taxes are required (1989, 52).

Among the policy goals of the preservation forces have been enactment of a wilderness bill, establishment of

> seashore and shoreline areas...to improve both the quality and quantity of public recreation opportunities; [the formulation of] a comprehensive federal recreation lands program;...surveying where additional national parks, forest, and seashores should be proposed;...ensuring that sufficient land be acquired around federally financed reservoirs for recreation purposes; and...[the] establish[ment] [of] a long-range program for planning and providing open spaces for recreational facilities in urban areas (1989, 28).

[18] While many American Indian nations are currently struggling with environmental issues, as are state and local governments, we must be cautious about generalizing from the pop image of Indian-as-environmentalist created by the media. Karen was one of several of the New Mexico leaders concerned with the stereotyping of Indian people.

> There's almost a stereotype that Native Americans are the first environmentalists and yet, when I drive down my road at home it's my people that are throwing out their McDonalds' bags, that are throwing out their beer cans, that are throwing out their trash. Sure there's [an environmental] aspect to our religion and, for the most part, that influences people. They understand that, they know it, and they love it. But when you're away from that tradition, it stops (Karen, activist).

NOTES TO CHAPTER 6

[1]Olson identifies an exception to his model, the "privileged group." If the group that would benefit from collective action is sufficiently small and the cost-benefit ratio of collective action for the group sufficiently favorable, there may be calculated individual action in the collective interest, even without selective incentives.

[2]Research on the political activism of women of color supports this notion. For example, Naples found that despite the fact that "community work was low paid, offered little opportunity for advancement, and was highly unstable," many of the women she studied who were active in community service remained involved for long periods and continued to volunteer after retirement from their paid positions (1992, 459. Also see Pardo 1998; Kaplan 1997; Hoikkala 1995; Sen 1995; Gilkes 1994; Hardy-Fanta 1993; Ford 1990; Kingsolver 1989).

[3]In a 1992 study of Latina and African American women's community work, Naples similarly found that racial discrimination was an important theme and motivating factor for much of the women's grassroots activism. For more on the New Mexico leaders' responses to racism, see chapter 5.

[4]This is not to say that material benefits may not accrue for neighborhoods or communities. On the contrary, the motivation for many of the New Mexico leaders was to improve the quality of life for their communities through physical improvements to infrastructure, and the provision of social services to needy populations. I am simply pointing out that the leaders themselves did not seek economic or monetary pay-offs (selective material incentives) for their political involvement.

[5]In her essay on engendering citizenship, Longo notes the nexus between feminist ideology and the values inherent in (participatory) democratic politics.

> Recent studies of the gender gap have shown that a feminist identity does foster the expression of a woman's perspective. Findings show that the gap is not so much between men and women as it is between men and feminist women on issues ranging from domestic issues to foreign policy issues. Such distinctive policy preferences of feminists do not stem solely from a woman's perspective, but are shaped by a commitment to democratic values. Thus only when the two are combined do we arrive at a distinctive set of policy preferences which have the potential to alter substantially the political agenda (1997, 9).

[6]As Diamond and Hartsock point out, "much of what women want and need is not the same as what men want and need. Reproductive freedom and access to abortion are perhaps the most prominent examples" (1981, 719).

[7]This finding is echoed by Naples, who studied the community work of Latinas and African American women in working class neighborhoods. She coined the term "activist mothering" to describe their particular brand of grassroots politics - a blending of "mothering practices, political activism, and community work" - undertaken to meet the needs of their children and community (1992, 448). More importantly, Naples notes that

> once they became active in community work, [these women's] experiences and acts of resistance defied the dominant definition of *motherhood* as emphasizing work performed in the private sphere of the fam-

ily or in face-to-face interaction with those in need (1992: 449. Italics in the original).

Indeed, their struggles for social and economic justice, and for racial and gender equality placed these "activist mothers" squarely in the realm of politics.

[8]One of the women who lobbied for the teaching of Diné (Navajo) in the Albuquerque public schools was a grassroots activist who also served on the board of the Parent Teacher Organization at her child's school.

[9]At first glance, 60 percent may seem like a surprisingly low figure considering that the sample of New Mexico leaders was chosen to reflect women active in environmental policymaking. However, these leaders were chosen not because they were "environmentalists" but because of their support of environmental justice, which reflects a broader social agenda of economic justice and racial and gender equity. In other words, the New Mexico leaders did not limit themselves to environmental issues but instead incorporated a broad spectrum of policy goals designed to improve the quality of life for their communities. (The critical distinctions between the grassroots environmental justice movement and the national environmental movement are discussed in greater detail in chapter 3.)

[10]The highly controversial Waste Isolation Pilot Project (WIPP), operated under the auspices of the U.S. Department of Energy, is an underground facility located in southeastern New Mexico designed to store some of the nation's medium- and low level radioactive waste.

[11]The Resource Conservation and Recovery Act (RCRA) regulates substances that "may cause, or significantly contribute to an increase in serious, irreversible, or incapacitating reversible illness; or pose a...hazard to human health or the environment" to a standard "necessary to protect human health and the environment."

[12]Sapiro identifies some of the same strategies used by the women's movement and by women's organizations. However, she classifies these strategies as one of seven types: conventional pressure techniques, nonconventional pressure techniques, consciousness-raising (empowerment), creating alternative institutions, separatism, caucusing, and international organization (1981, 709). For a marvelously detailed discussion of the multitude of strategies used by women around the world and through history to effect political change, see B. A. Carroll (1989).

[13]Another study of women in government similarly reported that "women were more likely than men...to value public input and participation in administrative processes, rejecting the belief that administrative policy makers could fare well without citizen or clientele participation" (Antolini 1984 quoting Neuse, 25-26).

[14]These findings parallel those reported in a growing body of scholarship that examines the politics of American Indian women and Hispanas. See for example, Naples (1998b), Pardo (1998, 1995, and 1990), Prindeville and Bretting (1998), Mack-Canty (1997), Hoikkala (1995), Sen (1995), Chiste (1994), Cruz Takash (1993), Hardy-Fanta (1993), Jaimes (1992), McCoy (1992), Bonilla-Santiago (1991), Ford (1990), Kingsolver (1989), Zavella (1988), Coyle, Hershatter, and Honig (1980).

NOTES TO CHAPTER 7

[1]I am familiar with only one empirical study that examined Chicana's support for, and conceptualizations of, feminism (see Pesquera and Segura 1993). While numerous scholars have argued for alternative feminisms to better reflect the perspectives, needs, and experiences of women of color (see for example Basu 1995; Johnson-Odim 1991; Chai and De Cambra 1989; hooks 1981; Cotera 1980), they have not conducted empirical studies. Where Warren (1987) suggested alternative approaches to both environmentalism and feminism following her examination of the connections among feminism, women of color, and environmentalism, she did not conduct empirical research either.

[2]U.S. Department of Commerce. *1990 Census of Population - General Population Characteristics, New Mexico*. 1990 CP-1-33 (from Table 28).

[3]This figure reflects the number of women registered to vote in the state in 1995 (Robertson 1995).

References

Ackelsberg, Martha A. 1988. Communities, Resistance, and Women's Activism: Some Implications for a Democratic Polity. In *Women and the Politics of Empowerment*, ed. A. Bookman and S. Morgen, 297–313. Philadelphia: Temple University Press.

———. 1984. Women's Collaborative Activities and City Life: Politics and Policy. In *Political Women: Current Roles in State and Local Government*, ed. J. A. Flammang, 242–259. Beverly Hills: Sage.

Acker, Joan, Kate Barry, and Johanna Esseveld. 1991. Objectivity and Truth: Problems in Doing Feminist Research. In *Beyond Methodology: Feminist Scholarship as Lived Research,* ed. M. M. Fonow and J. A. Cook, 133–153. Bloomington: Indiana University Press.

Acosta-Belén, Edna. 1993. Defining a Common Ground: The Theoretical Meeting of Women's, Ethnic, and Area Studies. In *Researching Women in Latin America and the Caribbean*, ed. E. Acosta-Belén and C. E. Bose, 175–186. Boulder: Westview Press.

Adeola, Francis O. 1994. Environmental Hazards, Health, and Racial Inequity in Hazardous Waste Distribution. *Environment and Behavior* 26(1): 99–126.

Afshar, Haleh. 1996. Introduction. In *Women and Politics in the Third World*, ed. H. Afshar, 1–6. New York: Routledge.

Aks, Judith H. 1996. Santa Clara Pueblo v. Martinez: Domination and Resistance at the Intersection of Race and Gender. Paper presented at the annual meetings of the Western Political Science Association in San Fransisco, CA, March 14–16.

Albers, Patricia. 1989. From Illusion to Illumination: Anthropological Studies of American Indian Women. In *Gender and Anthropology: Critical Reviews for*

Research and Teaching, ed. S. Morgen, 132–170. Washington, D.C.: American Anthropological Association.

———. 1983. Introduction: New Perspectives on Plains Indian Women. In *The Hidden Half: Studies of Plains Indian Women*, ed. P. Albers and B. Medicine, 1–26. Lanham: University Press of America.

Albrecht, Lisa, and Rose M. Brewer. 1990. Bridges of Power: Women's Multicultural Alliances for Social Change. In *Bridges of Power: Women's Multicultural Alliances*, ed. L. Albrecht and R. M. Brewer, 2–22. Philadelphia: New Society.

Allen, David W., James P. Lester, and Kelly M. Hill. 1995. Prejudice, Profits and Power: Assessing the Eco-Racism Thesis at the County Level. Paper presented at the annual meetings of the Western Political Science Association in Portland, OR, March 16–18.

Alston, Dana. 1990. Taking Back Our Lives: A Report to the Panos Institute on Environment, Community Development and Race in the United States. Washington, D.C.: The Panos Institute. Duplicated.

Ambler, Marjane. 1991. On the Reservations: No Haste, No Waste. *Planning* 57(11): 26–29.

Anaya, Toney, and Benjamin F. Chavis, Jr. People of Color Unite to Combat Environmental Racism. *Albuquerque Journal*, October 25.

Anderson, Kathryn, and Dana C. Jack. 1991. Learning to Listen: Interview Techniques and Analyses. In *Women's Words: The Feminist Practice of Oral History*, ed. S. Berger Gluck and D. Patai, 11–26. New York: Routledge.

Anderson, Kathryn, Susan Armitage, Dana Jack, and Judith Wittner. 1990. Beginning Where We Are: Feminist Methodology in Oral History. In *Feminist Research Methods*, ed. J. McCarl Nielsen, 94–112. Boulder: University of Colorado.

Antolini, Denise. 1984. Women in Local Government: An Overview. *Political Women: Current Roles in State and Local Government,* ed. J. A. Flammang, 23–40. Beverly Hills: Sage.

Aragón de Valdez, Theresa. 1980. Organizing as a Political Tool for the Chicana. *Frontiers* 5(2): 7–13.

Austin, Regina, and Michael Schill. 1994. Black, Brown, Red and Poisoned. *The Humanist* 54(4): 9–16.

Baca Zinn, Maxine and Bonnie Thornton Dill. 1994. Difference and Domination. In *Women of Color in U.S. Society*, ed. M. Baca Zinn and B. Thornton Dill, 3–12. Philadelphia: Temple University Press.

Baldassare, Mark, and Cheryl Katz. 1992. The Personal Threat of Environmental Problems as a Predictor of Environmental Practices. *Environment and Behavior* 24(5): 602–616.

Basu, Amrita. 1995. Introduction. In *The Challenge of Local Feminisms: Women's Movements in Global Perspective,* ed. A. Basu, 1–21. Boulder: Westview Press.

Bataille, Gretchen M., and Kathleen Mullen Sands. 1984. *American Indian Women: Telling Their Lives*. Lincoln: University of Nebraska Press.

Baumgartner, Frank R., and Bryan D. Jones. 1993. *Agendas and Instability in American Politics*. Chicago: University of Chicago Press.

Bayes, Jane H. 1982. *Minority Politics and Ideologies in the United States*. Novato: Chandler & Sharp Publishers.

Bellows, Anne C. 1996. Where Kitchen and Laboratory Meet: The 'Tested Food for Silesia' Program. In *Feminist Political Ecology: Global Issues and Local Experiences,* ed. D. Rocheleau, B. Thomas-Slayter, and E. Wangari, 251–270. New York: Routledge.

Belmonte, Bianca. 1996. Hispanic Women of the New Mexico State Legislature, 1922–1940s: Feminists in Politician's Clothes. Unpublished manuscript, University of New Mexico.

Benmayor, Rina. 1991. Testimony, Action Research, and Empowerment: Puerto Rican Women and Popular Education. In *Women's Words: The Feminist Practice of Oral History,* ed. S. Berger Gluck and D. Patai, 159–174. New York: Routledge.

Bennett, Stephen Earl, and Linda L. M. Bennett. 1986. Political Participation. *Annual Review of Political Science*: 157–204.

Berger Gluck, Sherna. 1991. Advocacy Oral History: Palestinian Women in Resistance. In *Women's Words: The Feminist Practice of Oral History*, ed. S. Berger Gluck and D. Patai, 205–220. New York: Routledge.

Berger Gluck, Sherna, and Daphne Patai, eds. 1991. *Women's Words: The Feminist Practice of Oral History*. New York: Routledge.

Blauner, Bob. 1987. Colonized and Immigrant Minorities. In *From Different Shores: Perspectives on Race and Ethnicity in America*, ed. R. Takaki, 149–160. New York: Oxford University Press.

Boneparth, Ellen. 1984. Resources and Constraints on Women in the Policymaking Process: State and Local Arenas. In *Political Women: Current Roles in State and Local Government,* ed. J. A. Flammang, 277–290. Beverly Hills: Sage.

Boneparth, Ellen, and Emily Stoper. 1988. Introduction: A Framework for Policy Analysis. In *Women, Power and Policy,* ed. E. Boneparth and E. Stoper, 1–19. New York: Pergamon Press.

Bonilla-Santiago, Gloria. 1991. Hispanic Women Breaking New Ground Through Leadership. *Latino Studies Journal* 2:19–37.

Bonvillain, Nancy. 1989. Gender Relations in Native North America. *American Indian Culture and Research Journal* 13(2): 1–28.

Bookman, Ann, and Sandra Morgen, eds. 1988. *Women and the Politics of Empowerment*. Philadelphia: Temple University Press.

Borland, Katherine. 1991. 'That's Not What I Said:' Interpretive Conflict in Oral Narrative Research. In *Women's Words: The Feminist Practice of Oral History,* ed. S. Berger Gluck and D. Patai, 63–76. New York: Routledge.

Bretting, John G., and Diane-Michele Prindeville. 1998. Environmental Justice and the Role of Indigenous Women Organizing Their Communities. In *Race and the Environment*, ed. D. Camacho, 141–164. Durham: Duke University Press.

Brown, Dee. 1970. *Bury My Heart At Wounded Knee: An Indian History of the American West*. London: Pan Books Ltd.

Brú-Bistuer, Josepa. 1996. Spanish Women Against Industrial Waste: A Gender Perspective on Environmental Grassroots Movements. In *Feminist Political Ecology: Global Issues and Local Experiences,* ed. D. Rocheleau, B. Thomas-Slayter, and E. Wangari, 105–126. New York: Routledge.

Brugge, David M. 1983. Navajo Prehistory and History to 1850. In *Handbook of North American Indians: Southwest* vol.10, ed. A. Ortiz, 489–501. Washington, D.C.: Smithsonian Institution.

Bryant, Bunyan, and Paul Mohai, eds. 1990. *The Proceedings of the Michigan Conference on Race and the Incidence of Environmental Hazards*. Ann Arbor: University of Michigan School of Natural Resources.

Bryant, Pat. 1991. Fertile Ground Between Environmentalists and People of Color. *Other Voices* 1(1): 10–11.

———. 1989. Toxics and Racial Justice. *Social Policy* 20(1): 48–52.

Buker, Eloise. 1987. Storytelling Power: Personal Narratives and Political Analysis. *Women & Politics* 7(3): 29–46.

Bullard, Robert D. 1995. Decision Making. In *Faces of Environmental Racism: Confronting Issues of Global Justice,* ed. L. Westra and P. S. Wenz, 3–28. Lanhman: Rowman and Littlefield Publishers.

———. 1993. Anatomy of Environmental Racism and the Environmental Justice Movement. In *Confronting Environmental Racism: Voices from the Grassroots,* ed. R. D. Bullard, 15–39. Boston: South End Press.

———. 1990. Environmental Blackmail in Minority Communities. In *The Proceedings of the Michigan Conference on Race and the Incidence of Environmental Hazards,* ed. B. Bryant and P. Mohai, 60–75. Ann Arbor: University of Michigan School of Natural Resources.

Bullard, Robert D., and Beverly Hendrix Wright. 1987. Environmentalism and the Politics of Equity: Emergent Trends in the Black Community. *Mid-American Review of Sociology* 12(2): 21–37.

Burrell, Barbara C. 1994. *A Woman's Place is in the House: Campaigning for Congress in the Feminist Era*. Ann Arbor: University of Michigan Press.

Bystydzienski, Jill M. 1992a. Introduction. In *Women Transforming Politics: Worldwide Strategies for Empowerment,* ed. J. M. Bystydzienski, 1–8. Indianapolis: Indiana University Press.

———. 1992b. Conclusion. In *Women Transforming Politics: Worldwide Strategies for Empowerment*, ed. J. M. Bystydzienski, 203–212. Indianapolis: Indiana University Press.

Cable, Sherry. 1992. Women's Social Movement Involvement: The Role of Structural Availability in Recruitment and Participation Processes. *The Sociological Quarterly* 33(1): 35–50.

Cable, Sherry, and Michael Benson. 1993. Acting Locally: Environmental Injustice and the Emergence of Grass-roots Environmental Organizations. *Social Problems* 40(4): 464–477.

Cable, Sherry, and Charles Cable. 1995. *Environmental Problems, Grassroots Solutions: The Politics of Grassroots Environmental Conflict.* New York: St. Martin's Press.

Campbell, Connie. 1996. Out On the Front Lines But Still Struggling for Voice: Women in the Rubber Tappers' Defense of the Forest in Xapuri, Brazil. In *Feminist Political Ecology: Global Issues and Local Experiences,* ed. D. Rocheleau, B. Thomas-Slayter, and E. Wangari, 27–61. New York: Routledge.

Campos Carr, Irene. 1988–89. A Survey of Selected Literature on La Chicana. *NWSA Journal* 1(2): 253–273.

Cannon, Joan Bartczak, and Mary Ann Marusich-Smith. 1986. Political Marriages: When the Wife is the Politician. *Women & Politics* 6(3): 57–74.

Cantor, Dorothy W., and Toni Bernay. 1992. *Women in Power: The Secrets of Leadership.* Boston: Houghton Mifflin Company.

Carmines, Edward G., and James A. Stimson. 1986. On the Structure and Sequence of Issue Evolution. *American Political Science Review* 80(3): 901–920.

Carpini, Michael X. Delli, and Ester R. Fuchs. 1993. The Year of the Woman? Candidates, Voters, and the 1992 Elections. *Political Science Quarterly* 108(1): 29–37.

Carroll, Berenice, A. 1989. 'Women Take Action!' Women's Direct Action and Social Change. *Women's Studies International Forum* 12(1): 3–24.

Carroll, Susan J. 1994. *Women as Candidates in American Politics*, 2nd ed. Indianapolis: Indiana University Press.

——— 1992. Women State Legislators, Womens' Organizations, and the Representation of Women's Culture in the United States. In *Women Transforming Politics: Worldwide Strategies for Empowerment,* ed. J. M. Bystydzienski, 24–40. Indianapolis: Indiana University Press.

———. 1991. Taking the Lead. *Journal of State Government* 64(2): 43–47.

———. 1989a. The Personal is Political: The Intersection of Private Lives and Public Roles Among Women and Men in Elected and Appointive Office. *Women & Politics* 9(2): 51–67.

———. 1989b. Gender Politics and the Socializing Impact of the Women's Movement. *Political Learning in Adulthood*, ed. R.S. Sigel, 306–339. Chicago: University of Chicago Press.

———. 1984. Women Candidates and Support for Feminist Concerns: The Closet Feminist Syndrome. *Western Political Quarterly* 37 (June): 307–323.

Carter, Clay. 1990. Race and Environmental Organizing: Pulling Down the Barriers. *Everyones' Back Yard* (March/April): 6–8.

Caulfield, Henry P. 1989. The Conservation and Environmental Movements: An Historical Analysis. In *Environmental Politics and Policy: Theories and Evidence*, ed. J. P. Lester, 13–56. Durham: Duke University Press.

Chai, Alice Yun, and Ho'oipo De Cambra. 1989. Evolution of Global Feminism Through Hawaiian Feminist Politics. *Women's Studies International Forum* 12(1): 59–64.

Chapman, Jenny. 1987. Adult Socialization and Out-Group Politicization: An Empirical Study of Consciousness-Raising. *British Journal of Political Science* 17: 315–340.

Chavez, Fray Angelico. 1979. Genízaros. In *Handbook of North American Indians: Southwest* vol.9, ed. A. Ortiz, 198–200. Washington, D.C.: Smithsonian Institution.

Chavis, Benjamin, Jr. 1993. Foreword. In *Confronting Environmental Racism: Voices from the Grassroots*, ed. R. D. Bullard, 3–5. Boston: South End Press.

Chiste, Katherine Beaty. 1994. Aboriginal Women and Self-government: Challenging Leviathan. *American Indian Culture and Research Journal* 18(3): 19–43.

Chowdhury, Najma, and Barbara J. Nelson. 1994a. Redefining Politics: Patterns of Women's Political Engagement from a Global Perspective. In *Women and Politics Worldwide,* ed. B. J. Nelson and N. Chowdhury, 3–24. New Haven: Yale University Press.

———. 1994b. Research Design and Practice: Methodological Issues in Feminist Comparative Politics Research. In *Women and Politics Worldwide*, ed. B. J. Nelson and N. Chowdhury, 25–40. New Haven: Yale University Press.

Churchill, Ward. 1992. The Earth is Our Mother: Struggles for American Indian Land and Liberation in the Contemporary United States. In *The State of Native America: Genocide, Colonization, and Resistance*, ed. M. A. Jaimes, 139–188. Boston: South End Press.

Churchill, Ward, and Glenn T. Morris. 1992. Table: Key Indian Laws and Cases. In *The State of Native America: Genocide, Colonization, and Resistance*, ed. M. A. Jaimes, 13–21. Boston: South End Press.

Citizen's Clearinghouse for Hazardous Wastes, Inc. 1992. *Everyone's Backyard* 10(1): Special Edition: CCHW's 1991 Yearbook.

———. 1990. Environmental Groups Challenged to Integrate. *Everyone's Backyard* (March-April): 8.

Clark, Janet, and Cal Clark. 1984. The Growth of Women's Candidacies for Nontraditional Political Offices in New Mexico. *Social Science Journal* 21(1): 57–65.

Cohen, Cathy J., Kathleen B. Jones, and Joan C. Tronto. 1997. Introduction: Women Transforming U.S. Politics: Sites of Power/Resistance. In *Women*

Transforming Politics: An Alternative Reader, ed. C. J. Cohen, K. B. Jones, and J. C. Tronto, 1–12. New York: New York University Press.

Conrad, Claire. 1991. Lori Goodman: On-the-Job Training. *Voces Unidas* 1(4): 12 (2). Constitution of the State of New Mexico. 1995. Santa Fe: Office of the Secretary of State.

Contreras, Roberto. 1993a. Toxic Survivors Seek Protection. *Voces Unidas* 3(2): 7.

———. 1993b. Standing Her Ground. *Voces Unidas* 3(2): 9 (2).

Contreras, Ruth, and Rose Shaw. 1993. Isleta Pueblo's First Woman Governor. *Voces Unidas* 3(2): 10 (2).

Cook, Judith A., and Mary Margaret Fonow. 1990. Knowledge and Women's Interests: Issues of Epistemology and Methodology in Feminist Sociological Research. In *Feminist Research Methods: Exemplary Readings in the Social Sciences*, ed. J. McCarl Nielsen, 69–93. Boulder: Westview.

Costain, Anne N. 1981. Representing Women: The Transition from Social Movement to Interest Group. *The Western Political Quarterly* 34(1): 100–113.

Cotera, Marta. 1980. Feminism: The Chicana and Anglo Versions. In *Twice a Minority: Mexican American Women*, ed. M. Melville, 217–234. St. Louis: C. V. Mosby Co.

Cotgrove, Stephen, and Andrew Duff. 1980. Environmentalism, Middle-class Radicalism and Politics. *Sociological Review* 28(2): 333–351.

Coyle, Laurie, Gail Hershatter, and Emily Honig. 1980. Women at Farah: An Unfinished Story. In *Mexican Women in the United States: Struggles Past and Present*, ed. M. Mora and A. R. Del Castillo, 117–143. Los Angeles: Chicano Studies Research Center Publications.

Crenson, Mathew A. 1971. *The Un-Politics of Air Pollution: A Study of Non-Decisionmaking in the Cities*. Baltimore: Johns Hopkins University Press.

Cruz Takash, Paule. 1993. Breaking Barriers to Representation: Chicana/Latina Elected Officials in California. *Urban Anthropology* 22(3–4): 325–360.

Cutter, Susan Caris. 1981. Community Concern for Pollution: Social and Environmental Influences. *Environment and Behavior* 13(1): 105–124.

Dabrowski, Irene. 1987. The Unnamed Political Woman. In *Women Leaders in Contemporary U.S. Politics*, ed. F. P. Le Veness and J. P. Sweeney, 137–149. Boulder: Lynne Rienner.

———. 1983. Working-Class Women and Civic Action: A Case Study of an Innovative Community Role. *Policy Studies Journal* 11(3): 427–435.

Darcy, R., Susan Welch, and Janet Clark. 1987. *Women, Elections, and Representation*. New York: Longman.

De la Garza, Rodolfo O., and David Vaughan. 1985. The Political Socialization of Chicano Elites: A Generational Approach. In *The Mexican American Experience: An Interdisciplinary Anthology*, ed. R. O. De la Garza, F. D. Bean, C. M. Bonjean, R. Romo, and R. Alvarez, 241–258. Austin: University of Texas Press.

Delamotte, Eugenia, Natania Meeker, and Jean O'Barr, eds. 1997. *Women Imagine Change: A Global Anthology of Women's Resistance from 600 B.C.E. to Present*. New York: Routledge.

Deutchman, Iva Ellen. 1992. Ungendered but Equal: Male Attitudes Toward Women in State Legislatures. *Polity* 24(3): 417–432.

———. 1986. Socialization to Power: Questions About Women and Politics. *Gender and Socialization to Power and Politics*, ed. R. M. Kelly, 79–91. New York: Haworth Press.

Devens, Carol. 1992. *Countering Colonization: Native American Women and Great Lakes Missions, 1630–1900*. Berkeley: University of California Press.

Dexter, Lewis. 1975. *Elite and Specialized Interviewing*. Evanston: Northwestern University Press.

Di Chiro, Giovanna. 1992. Defining Environmental Justice: Women's Voices and Grassroots Politics. *Socialist Review* 22(4): 93–130.

Diamond, Irene, and Nancy Hartsock. 1981. Beyond Interests in Politics: A Comment on Virginia Sapiro's 'When Are Interests Interesting? The Problem of Political Representation of Women.' *APSR* 75: 717–721.

Diane, Susan. 1995. Mescalero Apaches Continue the Fight Against the Storage of High Level Nuclear Waste on their Sacred Lands. *Enchanted Times* (Spring): 12–13.

Dietz, Mary G. 1985. Citizenship with a Feminist Face: The Problem with Maternal Thinking. *Political Theory* 13(1): 19–37.

Dodson, Debra, Susan J. Carroll, Ruth B. Mandel, Katherine E. Kleeman, Ronnee Schreiber, and Debra Liebowitz. 1995. *Voices, Views, Votes: The Impact of Women in the 103rd Congress*. Newark: Center for the American Woman and Politics, Rutgers University.

Downs, Anthony. 1972. Up and Down with Ecology - The 'Issue-Attention Cycle.' *The Public Interest* 28: 38–50.

———. 1957. *An Economic Theory of Democracy*. New York: Harper.

Duerst-Lahti, Georgia, and Rita Mae Kelly. 1995. On Governance, Leadership, and Gender. In *Gender, Power, Leadership, and Governance*, ed. G. Duerst-Lahti and R. M. Kelly, 11–37. Ann Arbor: University of Michigan Press.

Dunlap, Riley E. 1989. Public Opinion and Environmental Policy. In *Environmental Politics and Policy*, ed. J. P. Lester, 87–134. Durham: Duke University Press.

Eckstein, Harry. 1975. Case Study and Theory in Political Science. In *Handbook of Political Science*, vol.7, F. I. Greenstein and N. W. Polsby, 79–137. Reading: Addison-Wesley Publishing.

Eckstein, Susan. 1989. Power and Popular Protest in Latin America. In *Power and Popular Protest: Latin American Social Movements*, ed. S. Eckstein, 1–61. Berkeley: University of California Press.

Edelstein, Michael. 1988. *Contaminated Communities: The Social and Psychological Impacts of Residential Toxic Exposure*. Boulder: Westview Press.

Eggan, Fred. 1983. Comparative Social Organization. In *Handbook of North American Indians: Southwest* vol.10, ed. A. Ortiz, 723–742. Washington, D.C.: Smithsonian Institution.

———. 1979. Pueblos: Introduction. In *Handbook of North American Indians: Southwest* vol.9, ed. A. Ortiz, 224–235. Washington, D.C.: Smithsonian Institution.

Elassar, Nan, Kyle Mackenzie, and Yvonne Tixlery Vigil. 1980. *Las Mujeres: Conversations From a Hispanic Community*. New York: McGraw-Hill.

Elshtain, Jean Bethke. 1981. *Public Man, Private Woman: Women in Social and Political Thought*. Princeton: Princeton University Press.

Enslin, Tina. 1991. Keep Radioactive Waste Water Out of Sewers. *Albuquerque Journal*, 1 November.

Etienne, Mona, and Eleanor Leacock. 1980. Introduction: Women and Anthropology: Conceptual Problems. In *Women and Colonization: Anthropological Perspectives,* ed. M. Etienne and E. Leacock, 1–24. New York: Praeger.

Etter-Lewis, Gwendolyn. 1991. Black Women's Life Stories: Reclaiming Self in Narrative Texts. In *Women's Words: The Feminist Practice of Oral History,* ed. S. Berger Gluck and D. Patai, 43–58. New York: Routledge.

Evans, Sara M. 1993. Women's History and Political Theory: Toward a Feminist Approach to Public Life. In *Visible Women: New Essays on American Activism,* ed. N. A. Hewitt and S. Lebsock, 119–139. Chicago: University of Illinois Press.

Fahey, Maureen. 1978. Block by Block: Women in Community Organizing. *Women: A Journal of Liberation* 6(1): 24–29.

Faich, Ronald G., and Richard P. Gale. 1971. The Environmental Movement: From Recreation to Politics. *Pacific Sociological Review* 14(3): 270–287.

Feldman, Martha S. 1995. *Strategies for Interpreting Qualitative Data.* London: Sage.

Ferguson, Kathy. 1987. Male-Ordered Politics: Feminism and Political Science. In *Idioms of Inquiry: Critique and Renewal in Political Science,* ed. T. Ball, 209–229. New York: State University Press of New York.

Fernandez, Rita. 1993. Los Lunas Stops Waste Incinerator. *Voces Unidas* 3(3): 12.

Fernández-Kelly, M. Patricia, and Anna M. García. 1990. Power Surrendered, Power Restored: The Politics of Work and Family Among Hispanic Garment Workers in California and Florida. In *Women, Politics and Change*, ed. L. A. Tilly and P. Gurin, 130–149. New York: Russell Sage Foundation.

Ferree, Myra Marx, and Patricia Yancey Martin. 1995. Doing the Work of the Movement: Feminist Organizations. In *Feminist Organizations: Harvest of the Women's Movement,* ed. M. M. Ferree and P. Y. Martin, 3–23. Philadelphia: Temple University Press.

Fincher, Ruth, and Jacinta McQuillen. 1989. Women in Urban Social Movements. *Urban Geography* 10(6): 604–613.

Flammang, Janet A. 1984a. Introduction: A Reflection on the Themes of a 'Women's Politics.' In *Political Women: Current Roles in State and Local Government,* ed. J. A. Flammang, 9–20. Beverly Hills: Sage.

———. 1984b. Filling the Party Vacuum: Women at the Grassroots Level in Local Politics. *Political Women: Current Roles in State and Local Government,* ed. J. A. Flammang, 87–113. Beverly Hills: Sage Publications.

Fleck, John. 1995. Nuclear Waste Plan Rejected. *Albuquerque Journal,* 2 February.

Fonow, Mary Margaret, and Judith A. Cook, eds. 1991. *Beyond Methodology: Feminist Scholarship as Lived Research.* Indianapolis: Indiana University Press.

Ford, Ramona L. 1990. Native American Women Activists: Past and Present. Unpublished manuscript, Southwest Texas State University.

Forrest, Suzanne. 1989. *The Preservation of the Village: New Mexico's Hispanics and the New Deal.* Albuquerque: University of New Mexico.

Fowlkes, Diane L. 1992. *White Political Women: Paths from Privilege to Empowerment.* Knoxville: University of Tennessee Press.

Fox, Richard Logan. 1997. *Gender Dynamics in Congressional Elections.* London: Sage.

Frankovic, Kathleen A. 1982. Sex and Politics - New Alignments, Old Issues. *PS* 15: 439–448.

Freudenberg, Nicholas, and Carol Steinsapir. 1991. Not in Our Backyards: The Grassroots Environmental Movement. *Society and Natural Resources* 4: 235–245.

Freudenberg, Nicholas. 1984. *Not in Our Backyards! Community Action for Health and the Environment.* New York: Monthly Review Press.

Gale, Richard P. 1983. The Environmental Movement and the Left: Antagonists or Allies? *Sociological Inquiry* 53(2/3): 179–199.

Garcia, Alma M. 1989. The Development of Chicana Feminist Discourse, 1970–1980. *Gender and Society* 3: 217–238.

Garcia, F. Chris, Christine Marie Sierra, and Margaret Maier Murdock. 1991. The Politics of Women and Ethnic Minorities. In *Politics and Public Policy in the Contemporary American West,* ed. C. S. Thomas, 195–228. Albuquerque: University of New Mexico Press.

Garcia, F. Chris. 1994. To Get Along or To Go Along? Pluralistic Accommodation Versus Progress in New Mexico Politics. In *Contemporary New Mexico, 1940–1990,* ed. R. W. Etulain, 25–59. Albuquerque: University of New Mexico Press.

Gaylord, Clarice E., and Elizabeth Bell. 1995. Environmental Justice: A National Priority. In *Faces of Environmental Racism: Confronting Issues of Global*

Justice, ed. L. Westra and P. S. Wenz, 29–39. Lanhman: Rowman and Littlefield Publishers.

Gedicks, Al. 1993. *The New Resource Wars: Native and Environmental Struggles Against Multinational Corporations*. Boston: South End Press.

Genovese, Michael A. 1993. Women as National Leaders: What Do We Know? In *Women as National Leaders*, ed. M. A. Genovese, 211–217. London: Sage.

Genovese, Michael A., and Seth Thompson. 1993. Women as Chief Executives: Does Gender Matter? In *Women as National Leaders*, ed. M. A. Genovese, 1–12. London: Sage.

Gilkes, Cheryl Townsend. 1994. 'If It Wasn't for the Women...:' African American Women, Community Work, and Social Change. In *Women of Color in U.S. Society*, ed. M. Baca Zinn and B. Thornton Dill, 229–245. Philadelphia: Temple University Press.

Gilligan, Carol. 1987. In a Different Voice. In *Individualism and Commitment in American Life*, ed. R. N. Bellah, et al., 73–89. New York: Perennial Library.

Githens, Marianne. 1983. The Elusive Paradigm. Gender, Politics and Political Behavior: The State of the Art. In *Political Science: The State of the Discipline*, ed. A. W. Finifter, 471–499. Washington, D.C.: The American Political Science Association.

Githens, Marianne, and Jewel Prestage. 1978. Women State Legislators: Style and Priorities. *Policy Studies Journal* 7(2): 264–270.

Glover, Cindy. 1998. Women in New Mexico Politics. *Sage Magazine* (April): 8–11.

Goldman, Benjamin. 1991. *The Truth About Where You Live*. New York: Random House.

Gonzalez, Nancie L. 1969. *The Spanish Americans of New Mexico: A Heritage of Pride*. Albuquerque: University of New Mexico Press.

Gottlieb, Robert, and Helen Ingram. 1988. The New Environmentalists. *The Progressive* 52: 14–15.

Green, Rayna. 1980. Native American Women. *Signs: Journal of Women in Culture and Society* 6(2): 248–267.

Guerrero, Michael. 1994. Albuquerque Revisits Radioactive Dumping. *Voces Unidas* 4(2): 5 (2).

Guerrero, M., and L. Head. 1990. Informed Collective Action - A Powerful Weapon. In *Taking Back Our Lives*, ed. D. Alston, 24 (4). Washington, D.C.: The Panos Institute.

Gunn Allen, Paula. 1986. *The Sacred Hoop: Recovering the Feminine in American Indian Traditions*. Boston: Beacon Press.

Gurin, Patricia. 1985. Women's Gender Consciousness. *Public Opinion Quarterly* 49(2): 143–163.

Gutierrez, Ramon A. 1991. *When Jesus Came, the Corn Mothers Went Away*. Stanford: Stanford University Press.

Haggis, Jane. 1990. The Feminist Research Process - Defining a Topic. In *Feminist Praxis: Research, Theory and Epistemology in Feminist Sociology*, ed. L. Stanley, 67–79. New York: Routledge.

Hale, Sondra. 1991. Feminist Method, Process, and Self-Criticism: Interviewing Sudanese Women. In *Women's Words: The Feminist Practice of Oral History*, ed. S. Berger Gluck and D. Patai, 121–136. New York: Routledge.

Hall, Bob. 1988. Lessons from the Grassroots. In *Environmental Politics: Lessons from the Grassroots*, ed. B. Hall, 1–12. Durham: Institute for Southern Studies.

Hall, Nora. 1990. African-American Women Leaders and the Politics of Alliance Work. In *Bridges of Power: Women's Multicultural Alliances*, ed. L. Albrecht and R. M. Brewer, 74–94. Philadelphia: New Society Publishers.

Hamilton, Cynthia. 1993. Coping with Industrial Exploitation. In *Confronting Environmental Racism: Voices from the Grassroots*, ed. R. D. Bullard, 63–74. Boston: South End Press.

———. 1989. Women in Politics: Methods of Resistance and Change. *Women's Studies International Forum* 12(1): 129–135.

Hamilton, Lawrence C. 1985. Concern About Toxic Wastes: Three Demographic Predictors. *Sociological Perspectives* 28(4): 463–486.

Harding, Sandra. 1987. Introduction: Is There a Feminist Method? In *Feminism and Methodology*, ed. S. Harding, 1–14. Bloomington: Indian University Press.

Hardy-Fanta, Carol. 1993. *Latina Politics, Latino Politics: Gender, Culture, and Political Participation in Boston*. Philadelphia: Temple University Press.

Hartmann, Heidi I. 1987. The Family as the Locus of Gender, Class, and Political Struggle: The Example of Housework. In *Feminism and Methodology*, ed. S. Harding, 109–134. Bloomington: Indiana University Press.

Hartsock, Nancy C. M. 1987. The Feminist Standpoint: Developing the Ground for a Specifically Feminist Historical Materialism. In *Feminism and Methodology*, ed. S. Harding, 157–180. Bloomington: Indiana University Press.

Havens, Catherine M., and Lynne M. Healy. 1991. Do Women Make a Difference? *Journal of State Government* 64(2): 663–647.

Head, Louis. 1993. Take Back New Mexico! *Voces Unidas* 3(2): 1 (2).

Head, Louis and Michael Guerrero. 1991. Fighting Environmental Racism. *New Solutions* (Spring): 38–42.

Heiman, Michael. 1990. From 'Not in My Backyard' to 'Not in Anybody's Backyard:' Grassroots Challenge to Hazardous Waste Facility Siting. *Journal of the American Planning Association* 56(3): 359–362.

Hernandez, Patricia. 1980. Lives of Chicana Activists: The Chicano Student Movement (A Case Study). In *Mexican Women in the United States: Struggles Past and Present*, ed. M. Mora and A. Del Castillo, 17–25. Los Angeles: University of California Chicano Studies Research Center.

Hershey, Marjorie Randon, and David B. Hill. 1977–78. Is Pollution 'a White Thing'? Racial Differences in Pre-Adults' Attitudes. *Public Opinion Quarterly* 41(4): 439–458.

Hewitt, Nancy A. 1990. Varieties of Voluntarism: Class, Ethnicity and Women's Activism in Tampa. In *Women, Politics and Change*, ed. L. A. Tilly and P. Gurin, 63–86. New York: Russell Sage Foundation.

Hoikkala, Paivi Helena. 1995. Native American Women and Community Work in Phoenix, 1965–1980. [CD-ROM]. Abstract from: Dissertation Abstracts International, vol. 56–06A: 2382. Item No. AAI95–33382.

Holmes, Jack E. 1967. *Politics in New Mexico*. Albuquerque: University of New Mexico Press.

Holstein, James A., and Jaber F. Gubrium. 1995. *The Active Interview*. Beverly Hills: Sage.

Holtzman, Elizabeth, and Shirley Williams. 1989. Women in the Political World: Observations. In *Learning About Women: Gender, Politics, and Power*, ed. J. K. Conway, S. C. Bourque, and J. W. Scott, 25–33. Ann Arbor: University of Michigan Press.

hooks, bell 1981. *Ain't I a Woman: Black Women and Feminism*. Boston: South End Press.

Horton, John. 1992. The Politics of Diversity in Monterey Park, California. In *Structuring Diversity: Ethnographic Perspectives on the New Immigration*, ed. L. Lamphere, 215–245. Chicago: University of Chicago Press.

House, Glenda P. 1993. Testimonies Given by Impacted Communities. *Voces Unidas* 3(1): 1 (4).

Human Environment Center. 1991. Mission and Activities Statement, and program information. Washington, D.C.: April.

Hutchings, Vicky. 1994. Racism Plays a Key Role in Who Gets Toxic Waste Dumped on Them. *New Statesman and Society* (March): 31.

Ingram, Helen M., and Dean E. Mann. 1989. Interest Groups and Environmental Policy. In *Environmental Politics and Policy*, ed. J. P. Lester, 135–157. Durham: Duke University Press.

Inter-Hemispheric Education Resource Center. 1994. Concerned Citizens of Sunland Park. *BorderLines* 2(1): 4–5.

Iverson, Peter. 1983. The Emerging Navajo Nation. In *Handbook of North American Indians: Southwest* vol.10, ed. A. Ortiz, 736–740. Washington, D.C.: Smithsonian Institution.

Jackman, Robert W. 1985. Cross-National Statistical Research and the Study of Comparative Politics. *AJPS* 29(1): 161–182.

Jackson, Peter. 1993. Maine's Tribal Reps Hold Unique Status, But No Vote. *Sun-Journal*, 11 April.

Jacobs, Sue-Ellen. 1995. Continuity and Change in Gender Roles at San Juan Pueblo. In *Women and Power in Native North America*, ed. L. F. Klein and L. A. Ackerman, 177–213. Norman: University of Oklahoma Press.

Jacquez-Ortiz, Michele. 1995. Hispanic Women in Politics Doing More Than Treading Water. *La Herencia del Norte* (Spring): 30–31.

Jaimes, M. Annette. 1992. American Indian Women: At the Center of Indigenous Resistance in North America. In *The State of Native America: Genocide, Colonization, and Resistance*, ed. M. A. Jaimes, 311–344. Boston: South End Press.

Jankowski, Martin Sanchez. 1986. *City Bound: Urban Life and Political Attitudes Among Chicano Youth*. Albuquerque: University of New Mexico Press.

Jarvenpa, Robert. 1989. The Political Economy and Political Ethnicity of American Indian Adaptations and Identities. In *Ethnicity and Race in the USA: Toward the 21st Century*, ed. R. D. Alba, 29–48. New York: Routledge.

Joe, Jennie R., and Dorothy Lonewolf Miller. 1994. Cultural Survival and Contemporary American Indian Women in the City. In *Women of Color in U.S. Society*, ed. M. Baca Zinn and B. Thornton Dill, 185–202. Philadelphia: Temple University Press.

Johnson-Odim, Cheryl. 1991. Common Themes, Different Contexts: Third World Women and Feminism. In *Third World Women and the Politics of Feminism*, ed. C. T. Mohanty, A. Russo, and L. Torres, 314–327. Indianapolis: Indiana University Press.

Jones, Mary Lou. 1994. Missiles Over Dineh. *Voces Unidas* 4(1): 11.

Juarez, Teresa. Personal communication, 25 November 1991.

Junn, Jane. 1997. Assimilating or Coloring Participation? Gender, Race, and Democratic Political Participation. In *Women Transforming Politics: An Alternative Reader*, ed. C. J. Cohen, K. B. Jones, and J. C. Tronto, 387–397. New York: New York University Press.

Kahn, Kim Fridkin. 1996. *The Political Consequences of Being a Woman: How Stereotypes Influence the Conduct and Consequences of Political Campaigns*. New York: Columbia University Press.

Kaplan, Temma. 1997. *Crazy for Democracy: Women in Grassroots Movements*. New York: Routledge.

———. 1982. Female Consciousness and Collective Action: The Case of Barcelona, 1910–1918. *Signs: Journal of Women in Culture and Society* 7(3): 545–566.

Kathlene, Lyn. 1989. Uncovering the Political Impacts of Gender: An Exploratory Study. *Western Political Quarterly* 42(2): 397–421.

Katz, Jack. 1988. A Theory of Qualitative Methodology: The Social System of Analytic Fieldwork. In *Contemporary Field Research*, ed. R. M. Emerson, 127–148. Prospect Heights: Waveland Press.

Kelley, Matt. 1999. Crushing a Culture. *Albuquerque Journal*, 25 April.

Kelly, Rita Mae, Bernard Ronan, and Margaret E. Cawley. 1987. Liberal Positivistic Epistemology and Research on Women and Politics. *Women & Politics* 7(3): 11–27.

Kimball, Rene. 1991 Sandia Water Plan on Verge of Defeat. *Albuquerque Journal*, 5 November.

King, Lourdes Miranda. 1987. Puertorriqueñas in the United States: The Impact of Double Discrimination. In *From Different Shores: Perspectives on Race and Ethnicity in America*, ed. R. Takaki, 192–196. New York: Oxford University Press.

Kingsolver, Barbara. 1989. *Holding the Line: Women in the Great Arizona Mine Strike of 1983*. Ithaca: ILR Press.

Klein, Laura F., and Lillian A. Ackerman. 1995. Introduction. In *Women and Power in Native North America*, ed. L. F. Klein and L. A. Ackerman, 3–16. Norman: University of Oklahoma Press.

Knack, Martha C. 1995. The Dynamics of Southern Paiute Women's Roles. In *Women and Power in Native North America*, ed. L. F. Klein and L. A. Ackerman, 146–158. Norman: University of Oklahoma Press.

Koester, Susan H. 1988. 'By the Words of the Mouth Let Thee Be Judged:' The Alaska Native Sisterhood Speaks. *Journal of the West* 27(2): 35–44.

Krauss, Celene. 1998. Challenging Power: Toxic Waste Protests and the Politicization of White, Working-Class Women. In *Community Activism and Feminist Politics: Organizing Across Race, Class, and Gender*, ed. N. A. Naples, 129–150. New York: Routledge.

———. 1993. Women and Toxic Waste Protests: Race, Class and Gender as Resources of Resistance. *Qualitative Sociology* 16(3): 247–262.

Kutsche, Paul, and John R. Van Ness. 1981. *Cañones: Values, Crisis, and Survival in a Northern New Mexico Village*. Albuquerque: University of New Mexico Press.

Kutz, Jack. 1989. *Grassroots New Mexico: A History of Citizen Activism*. Albuquerque: The Inter-Hemispheric Education Resource Center.

LaDuke, Winona. 1991. Environmental Work: An Indigenous Perspective. *Northeast Indian Quarterly* (Winter): 16–19.

LaFramboise, Teresa, Annaliese Heyle, and Emily Ozer. 1990. Changing and Diverse Roles of Women in American Indian Cultures. *Sex Roles* 22(7/8): 455–476.

Lamphere, Louise, Patricia Zavella, and Felipe Gonzales. 1993. *Sunbelt Working Mothers: Reconciling Family and Factory*. Ithaca: Cornell University Press.

Lange, Charles H. 1979. Relations of the Southwest with the Plains and Great Basin. In *Handbook of North American Indians: Southwest* vol.9, ed. A. Ortiz, 201–205. Washington, D.C.: Smithsonian Institution.

Laws, Rufina Marie. 1995. Spiritual Conflict Underlies Nuclear Waste Controversy. *Albuquerque Journal*, 8 March.

Lawson, Bill. 1995. Living for the City: Urban United States and Environmental Justice. In *Faces of Environmental Racism: Confronting Issues of Global Justice*, ed. L. Westra and P. S. Wenz, 41–55. Lanhman: Rowman and Littlefield Publishers.

Lee, Charles. 1993. Beyond Toxic Wastes and Race. In *Confronting Environmental Racism: Voices from the Grassroots*, ed. R. D. Bullard, 41–52. Boston: South End Press.

———. 1990. Toxic Waste and Race in the United States. In *The Proceedings of the Michigan Conference on Race and the Incidence of Environmental Hazards*, ed. B. Bryant and P. Mohai, 6–27. Ann Arbor: University of Michigan, School of Natural Resources.

Lester, James P. 1989. Introduction. In *Environmental Politics and Policy*, ed. J. P. Lester, 1–9. Durham: Duke University Press.

Le Veness, Frank P., and Jane P. Sweeney, eds. 1987. *Women Leaders in Contemporary U.S. Politics*. Boulder: Lynne Rienner.

Liebowitz, Debra J., and Susan J. Carroll. 1996. *Women and American Politics: A Research Agenda for the 21st Century.* Newark: Center for the American Woman and Politics, Rutgers University.

Lijphart, Arend. 1975. Comparative Politics and the Comparative Method. *American Political Science Review* 65: 682–693.

Linthicum, Leslie. 1998. Woman's Election a Laguna 1st. *Albuquerque Journal*, 29 December.

———. 1996. Laguna May OK Women in Politics. *The Sunday Journal*, 1 December.

Lobao, Linda. 1993. Women in Revolution: The Mobilization of Latin American Women in Revolutionary Guerilla Movements. In *Feminist Frontiers III*, ed. L. Richardson and V. Taylor, 470–482. New York: McGraw Hill.

Lofland, John, and Lyn Lofland. 1995. *Analyzing Social Settings: A Guide to Qualitative Observation and Analysis*, 3rd ed. Belmont: Wadsworth Publishing.

Longo, Patrizia. 1997. Engendering Citizenship and Democracy. Paper presented at the annual meetings of the Western Political Science Association, Tucson, Arizona, March 13–15.

Lowe, George D., and Thomas K. Pinhey. 1982. Rural-Urban Differences in Support for Environmental Protection. *Rural Sociology* 47(1): 114–128.

Lowe, George D., Thomas K. Pinhey, and Michael D. Grimes. 1980. Public Support for Environmental Protection: New Evidence from National Surveys. *Pacific Sociological Review* 23(4): 423–445.

Lurie, Nancy Oestreich. 1972. Indian Women: A Legacy of Freedom. *Look at the Mountain*, ed. R. L. Iacopi, 29–36. San Jose: Gousha Publications.

Lynch, Robert N. 1986. Women in Northern Paiute Politics. *Signs: Journal of Women in Culture and Society* 11(21): 352–366.

Mack-Canty, Colleen. 1997. Women Political Activists: Taking the State Back Out. Paper presented at the Annual Meetings of the Western Political Science Association, Tucson, Arizona, March 13–15.

MacManus, Susan A., Charles S. Bullock III, and Barbara P. Grothe. 1986. A Longitudinal Examination of Political Participation Rates of Mexican American Females. *Social Science Quarterly* 67: 604–612.

MacManus, Susan A., and Carol A. Cassel. 1988. Mexican-Americans in City Politics: Participation, Representation and Political Preferences. *Latinos and the Political System*, ed. F. C. Garcia, 201–212. South Bend: University of Notre Dame Press.

Maltz, Daniel, and JoAllyn Archambault. 1995. Gender and Power in Native North America: Concluding Remarks. In *Women and Power in Native North America*, ed. L. F. Klein and L. A. Ackerman, 230–250. Norman: University of Oklahoma Press.

Marquez, Benjamin. 1995. Organizing Mexican-American Women in the Garment Industry: La Mujer Obrera. *Women & Politics* 15(1): 65–87.

———. 1994. The Politics of Environmental Justice in Mexican American Neighborhoods. Paper presented at the annual meetings of the Western Political Science Association in Albuquerque, NM, March 10–12.

Mayer, Laurence. 1989. *Redefining Comparative Politics: Promise Versus Performance*. Newbury Park: Sage.

McCarl Nielsen, Joyce. 1990. Introduction. In *Feminist Research Methods: Exemplary Readings in the Social Sciences*, ed. J. M. Nielsen, 1–37. Boulder: University of Colorado.

McClain, Paula D., and Joseph Stewart, Jr. 1995. *Can We All Get Along? Racial and Ethnic Minorities in American Politics*. Boulder: Westview Press.

McCoy, Melanie. 1992. Gender or Ethnicity: What Makes a Difference? A Study of Women Tribal Leaders. *Women & Politics* 12(3): 57–68.

McCraken, Grant. 1988. *The Long Interview*. Newbury Park: Sage Publications.

McGlen, Nancy E., and Karen O'Connor. 1995. *Women, Politics, and American Society*. Englewood Cliffs: Prentice-Hall.

Medicine, Bea. 1978. *The Native American Woman: A Perspective*. Austin: National Educational Laboratory Publishers.

Mehta, Manjari. 1996. 'Our Lives Are No Different From That of Our Buffaloes:' Agricultural Change and Gendered Spaces in a Central Himalayan Valley. In *Feminist Political Ecology: Global Issues and Local Experiences*, ed. D. Rocheleau, B. Thomas-Slayter, and E. Wangari, 180–210. New York: Routledge.

Mies, Maria. 1991. Women's Research or Feminist Research? The Debate Surrounding Feminist Science and Methodology. Translated by Andy Spencer. In *Beyond Methodology: Feminist Scholarship as Lived Research*, ed. M. M. Fonow and J. A. Cook, 60–84. Bloomington: Indian University Press.

Miles, Matthew B., and A. Michael Huberman. 1994. *Qualitative Data Analysis*, 2nd ed. London: Sage Publications.

Miller, Bruce G. 1992. Women and Politics: Comparative Evidence from the Northwest Coast. *Ethnology* 31(4): 367–375.

Miller, Vernice, Moya Hallstein, and Susan Quass. 1996. Feminist Politics and Environmental Justice: Women's Community Activism in West Harlem, New York. *Feminist Political Ecology: Global Issues and Local Experiences*, ed. D. Rocheleau, B. Thomas-Slayter, and E. Wangari, 62–86. New York: Routledge.

Millman, Marcia, and Rosabeth Moss Kanter, eds. 1975. *Another Voice: Feminist Perspectives on Social Life and Social Science*. Garden City: Doubleday.

Milroy, Beth Moore. 1991. Taking Stock of Planning, Space, and Gender. *Journal of Planning and Literature* 6(1): 3–15.

Minister, Kristina. 1991. A Feminist Frame for the Oral History Interview. In *Women's Words: The Feminist Practice of Oral History*, ed. S. Berger Gluck and D. Patai, 27– 41. New York: Routledge.

Mitchell, Robert Cameron. 1979. Silent Spring/Solid Majorities. *Public Opinion* 2(4): 16 (3).

Moe, Terry. 1981. Toward a Broader View of Interest Groups. *Journal of Politics* 4(3): 531–543.

Molyneaux, Maxine. 1985. Mobilization Without Emancipation? Women's Interests, the State, and Revolution in Nicaragua. *Feminist Studies* 11:227–254.

Moreno, Fernando. (Producer) *Environment, Race and Class: The Poisoning of Communities of Color.* Video documentary aired August 31, 1991, Community Cable Channel 27, Albuquerque, NM.

Morgen, Sandra. 1988. 'It's the Whole Power of the City Against Us!:' The Development of Political Consciousness in a Women's Health Care Coalition. In *Women and the Politics of Empowerment*, ed. A. Bookman and S. Morgen, 97–115. Philadelphia: Temple University Press.

Morgen, Sandra, and Ann Bookman. 1988. Rethinking Women and Politics: An Introductory Essay. In *Women and the Politics of Empowerment,* ed. A. Bookman and S. Morgen, 3–29. Philadelphia: Temple University Press.

Morris, Aldon D. 1984. *The Origins of the Civil Rights Movement: Black Communities Organizing for Change*. New York: The Free Press.

Morris K. Udall – A Lifetime of Service to Arizona and the United States, selected speeches. In *The American Indians and Civil Rights* [online] Available from <http://www.library.arizona.edu/branches/spc/udall/hidden.html>

Naples, Nancy A. 1998a. Women's Community Activism and Feminist Activist Research. In *Community Activism and Feminist Politics: Organizing Across Race, Class, and Gender,* ed. N. A. Naples, 1–27. New York: Routledge.

———. 1998b. Women's Community Activism: Exploring the Dynamics of Politicization and Diversity. In *Community Activism and Feminist Politics: Organizing Across Race, Class, and Gender*, ed. N. A. Naples, 327–349. New York: Routledge.

———. 1992. Activist Mothering: Cross-Generational Continuity in the Community Work of Women from Low-Income Urban Neighborhoods. *Gender and Society* (6)3: 441–463.

Naranjo, Loretta. 1994. Martineztown Defeats Courthouse! *Voces Unidas* 4(2): 4.

Neiman, Max, and Ronald O. Loveridge. 1981. Environmentalism and Local Growth Control: A Probe into the Class Bias Thesis. *Environment and Behavior* 13(6): 759– 772.

Nelson, Candace, and Marta Tienda. 1989. The Structuring of Hispanic Ethnicity: Historical and Contemporary Perspectives. In *Ethnicity and Race in the USA: Toward the 21st Century,* ed. R. D. Alba, 49–74. New York: Routledge.

New York Times News Service. 1991. Minorities Say No More Pollution in Back Yard. *Albuquerque Tribune*, 25 October.

Niethammer, Carolyn. 1977. *Daughters of the Earth: The Lives and Legends of American Indian Women.* New York: Collier MacMillan.

Olson, Mancur. 1965. *The Logic of Collective Action: Public Goods and the Theory of Groups.* Cambridge: Harvard University Press.

Omi, Michael, and Howard Winant. 1986. *Racial Formation in the United States: From the 1960s to the 1980s.* New York: Routledge.

O'Regan, Valerie. 1997. The Commitment of Female Policy-Makers to Women's Issues. Paper presented at the Annual Meetings of the Western Political Science Association, Tucson, Arizona, March 13–15.

Orleck, Annelise. 1997. Tradition Unbound: Radical Mothers in International Perspective. In *The Politics of Motherhood: Activist Voices from Left to Right,* ed. A. Jetter, A. Orleck, and D. Taylor, 3–11. Hanover: University Press of New England.

Ortiz, Alfonso. 1979. Introduction. In *Handbook of North American Indians: Southwest* vol.9, ed. A. Ortiz, 1–4. Washington, D.C.: Smithsonian Institution.

Orum, Anthony M. 1974. On Participation in Political Protest Movements. *Journal of Applied Behavioral Science* 10(2): 181–204.

Paehlke, Robert, and Pauline Vaillancourt Rosneau. 1993. Environment/Equity: Tensions in North American Politics. *Policy Studies Journal* 21(4): 672–686.

Pardo, Mary. 1998. Creating Community: Mexican American Women in Eastside Los Angeles. In *Community Activism and Feminist Politics: Organizing Across Race, Class, and Gender*, ed. N. A. Naples, 275–300. New York: Routledge.

———. 1995. Doing It For the Kids: Mexican American Activists, Border Feminists? *Feminist Organizations: Harvest of the Women's Movement*, ed. M. M. Ferree and P. Y. Martin, 356–371. Philadelphia: Temple University Press.

———. 1990. Mexican American Women Grassroots Community Activists: 'Mothers of East Los Angeles.' *Frontiers* 11(1): 1–7.

Parenti, Michael. 1970. Power and Pluralism: A View From the Bottom. *Journal of Politics* (August): 501–530.

———. 1967. Ethnic Politics and the Persistence of Ethnic Identification. *American Political Science Review* 61: 717–726.

Partners in the Environment. 1993. New Mexico: The Land of Enchantment? *The EAGLE* 1(6): 5 (2).

Patai, Daphne. 1991. U.S. Academics and Third World Women: Is Ethical Research Possible? In *Women's Words: The Feminist Practice of Oral History,* ed. S. Berger Gluck and D. Patai, 137–154. New York: Routledge.

Pateman, Carole. 1970. *Participation and Democratic Theory*. New York: Cambridge University Press.

Paul, Paula. 1995. Distaff Politics. *Sage Magazine* (September): 8–9.

Peña, Devon. 1992. The 'Brown' and the 'Green:' Chicanos and Environmental Politics in the Upper Rio Grande. *Capitalism Nature Socialism: A Journal of Socialist Ecology* 3(1): 79–103.

Peña, Devon, and Joseph Gallegos. 1993. Nature and Chicanos in Southern Colorado. In *Confronting Environmental Racism: Voices from the Grassroots,* ed. R. D. Bullard, 141–160. Boston: South End Press.

Perkins, Jerry. 1986. Political Ambition Among Black and White Women: An IntraGender Test of the Socialization Model. *Women & Politics* 6(1): 27–40.

Pesquera, Beatriz, and Denise Segura. 1993. There is No Going Back: Chicanas and Feminism. In *Chicana Critical Issues: Mujeres Activas en Letras y Cambio Social,* ed. N. Alarcón et al., 95–115. Berkeley: Third Woman Press.

Peterson, V. Spike, and Anne Sisson Runyan. 1993. *Global Gender Issues*. Boulder: Westview Press.

Phillips, Anne. 1991. *Engendering Democracy*. University Park: Pennsylvania State University Press.

Portney, Kent E. 1985. The Potential of the Theory of Compensation for Mitigation of Public Opposition to Hazardous Waste Siting: Some Evidence from Five Massachusetts Communities. *Policy Studies Journal* 14(1): 81–89.

Prindeville, Diane-Michele, and Teresa Braley Gomez. 1999. American Indian Women Leaders, Public Policy, and the Importance of Gender and Ethnic Identity. *Women & Politics* 20(2): 17–32.

Prindeville, Diane-Michele, and John G. Bretting. 1998. Indigenous Women Activists and Political Participation: The Case of Environmental Justice. *Women & Politics* 19(1): 39–58.

Prindeville, Diane-Michele, Felipe Gonzales, and Christine M. Sierra. June 1, 1992. The State of Ethnic and Race Relations in Albuquerque, New Mexico. Research Report for Project Change, Levi Strauss Foundation.

Pulido, Laura. 1993. Sustainable Development at Ganados del Valle. In *Confronting Environmental Racism: Voices from the Grassroots,* ed. R. D. Bullard, 123–139. Boston: South End Press.

Ragin, Charles C. 1987. *The Comparative Method: Moving Beyond Qualitative and Quantitative Strategies.* Berkeley: University of California Press.

Randall, Margaret. 1995. *Our Voices, Our Lives: Stories of Women from Central America and the Caribbean.* Monroe: Common Courage.

Rebolledo, Tey Diana, Erlinda Gonzales-Berry, and Millie Santillanes, eds. 1992. *Nuestras Mujeres: Hispanas of New Mexico, Their Images and Their Lives, 1582–1992.* Albuquerque: El Norte Publications/Academia.

Redhouse, John. 1984. The Seven Sisters: Native Women in Struggle. An unpublished memoir.

Reed, Adolph, Jr. 1988. The Black Urban Regime: Structural Origins and Constraints. In *Power, Community and the City*, ed. M. P. Smith, 138–175. New Brunswick: Transaction, Inc.

Remal, Gary J. 1991. Non-Voting Indian Representatives Play Unique Role in Maine Legislature. *Kennebec Journal*, 25 March.

Rendel, Margherita. 1981. Women, Power and Political Systems. *Women, Power and Political Systems,* ed. M. Rendel, 15–49. New York: St. Martin's Press.

Rhee, Suk. 1993. Southwest Network Campaign Update. *Voces Unidas* 3(1): 16 (2).

Robbins, Rebecca L. 1992. Self-determination and Subordination: The Past, Present, and Future of American Indian Governance. In *The State of Native America: Genocide, Colonization, and Resistance,* ed. M. A. Jaimes, 87–121. Boston: South End Press.

Robertson, John. 1995. Political Gains Come Slowly, Steadily for New Mexicans. *Albuquerque Journal*, 13 August.

Rocheleau, Dianne, Laurie Ross, and Julio Morrobel. 1996. From Forest Gardens to Tree Farms: Women, Men, and Timber in Zambrana-Chacuey, Dominican Republic. In *Feminist Political Ecology: Global Issues and Local Experiences,* ed. D. Rocheleau, B. Thomas-Slayter, and E. Wangari, 224–250. New York: Routledge.

Rocheleau, Dianne, Barbara Thomas-Slayter, and Esther Wangari. 1996. Gender and Environment: A Feminist Political Ecology Perspective. In *Feminist Political Ecology: Global Issues and Local Experiences*, ed. D. Rocheleau, B. Thomas-Slayter, and E. Wangari, 3–23. New York: Routledge.

Rodriguez, Sylvia. 1987. Land, Water, and Ethnic Identity in Taos. In *Land, Water, and Culture: New Perspectives on Hispanic Land Grants*, ed. C. L. Briggs and J. R. Van Ness, 313–403. Albuquerque: University of New Mexico Press.

Rodriguez-Trias, Helen. 1994. Women are Organizing: Environmental and Population Policies Will Never be the Same. *American Journal of Public Health* 84(9): 1379– 1382.

Roessel, Robert A. 1983. Navajo History, 1850–1923. In *Handbook of North American Indians: Southwest* vol.10, ed. A. Ortiz, 506–523. Washington, D.C.: Smithsonian Institution.

Romero, Fabi. 1997. Grupitos: Citizen Involvement at the Kitchen Table. *Nuestro Pueblo* (Summer): 7 (2).

Rose, Margaret. 1990. Traditional and Nontraditional Patterns of Female Activism in the United Farm Workers of America, 1962 to 1980. *Frontiers* 11(1): 26–32.

Rosenthal, Cindy Simon. 1995a. Once They Get There: The Role of Gender in Legislative Careers. *Extensions* (Spring): 15–17.

———. 1995b. The Role of Gender in Descriptive Representation. *Political Research Quarterly* 48(3): 599–611.

Ruddick, Sara. 1980. Maternal Thinking. *Feminist Studies* 6(2): 342–367.

Rule, Wilma. 1981. Why Women Don't Run: The Critical Contextual Factors in Women's Legislative Recruitment. *Western Political Quarterly* 34(1): 60–77.

Rusco, Elmer R., and Mary K. Rusco. 1986. Tribal Politics. In *Handbook of North American Indians: Great Basin* vol.11, ed. W. L. D'Azevedo, 558–572. Washington, D.C.: Smithsonian Institution.

Sabatier, Paul A., and Hank Jenkins-Smith. 1993. *Policy Change and Learning: An Advocacy Coalition Approach*. Boulder: Westview Press.

Sacks, Karen Brodkin. 1988. Gender and Grassroots Leadership. In *Women and the Politics of Empowerment*, ed. A. Bookman and S. Morgen, 77–94. Philadelphia: Temple University Press.

Salazar, Deborah, and Lisa A. Moulds. 1995. Toward an Integrated Politics of Social Justice and Environment: African American Leaders in Seattle. Paper presented at the annual meetings of the Western Political Science Association in Portland, OR, March 16–18.

Saint-Germain, Michelle A. 1989. Does Their Difference Make a Difference? The Impact of Women on Public Policy in the Arizona State Legislature. *Social Science Quarterly* 70(4): 956–969.

Sando, Joe. 1992. *Pueblo Nations: Eight Centuries of Pueblo Indian History*. Santa Fe: Clear Light Publishers.

Sapiro, Virginia. 1983. *The Political Integration of Women: Roles, Socialization, and Politics*. Urbana: University of Illinois Press.

———. 1981. Research Frontier Essay: When Are Interests Interesting? The Problem of Political Representation of Women. *American Political Science Review* 75:701–716.

Schwab, Jim. 1991. Blue Collar Groups are Saying, 'Not in Our Backyard'. *Planning* 57(10): 8–11.

Seager, Joni. 1996. 'Hysterical Housewives' and Other Mad Women: Grassroots Environmental Organizing in the United States. In *Feminist Political Ecology: Global Issues and Local Experiences*, ed. D. Rocheleau, B. Thomas-Slayter, and E. Wangari, 271–283. New York: Routledge.

Segura, Denise. 1994. Inside the Work Worlds of Chicana and Mexican Immigrant Women. In *Women of Color in U.S. Society*, ed. M. Baca Zinn and B. Thornton Dill, 95–111. Philadelphia: Temple University Press.

———. 1987. Labor Market Stratification: The Chicana Experience. In *From Different Shores: Perspectives on Race and Ethnicity in America*, ed. R. Takaki, 175–185. New York: Oxford University Press.

Seidman, Carrie. 1994. That Their Children May Inherit the Earth. *Sage Magazine* (October): 10–12.

Sen, Rinku. 1995. *We Are the Ones We Are Waiting For: Women of Color Organizing for Transformation*. Durham: United States Urban Rural Missions, World Council of Churches.

Seymour-Smith, Charlotte. 1986. *Dictionary of Anthropology*. Boston: G. K. Hall & Co.

Shapiro, Judith. 1986. Kinship. In *Handbook of North American Indians: Great Basin* vol.11, ed. W. L. D'Azevedo, 620–629. Washington, D.C.: Smithsonian Institution.

Shapiro, Robert Y., and Harpreet Mahajan. 1986. Gender Differences in Policy Preferences: A Summary of Trends from the 1960s to the 1980s. *Public Opinion Quarterly* (50)1: 42–61.

Shepardson, Mary. 1995. The Gender Status of Navajo Women. In *Women and Power in Native North America*, ed. L. F. Klein and L. A. Ackerman, 159–176. Norman: University of Oklahoma Press.

———. 1983. Development of Navajo Tribal Government. In *Handbook of North American Indians: Southwest* vol.10, ed. A. Ortiz, 624–635. Washington, D.C.: Smithsonian Institution.

Shields, M. Dale, Cornelia Butler Flora, Barbara Thomas-Slayter, and Gladys Buenavista. 1996. Developing and Dismantling Social Capital: Gender and Resource Management in the Philippines. In *Feminist Political Ecology: Global Issues and Local Experiences*, ed. D. Rocheleau, B. Thomas-Slayter, and E. Wangari, 155–179. New York: Routledge.

Sidel, Ruth. 1995. Toward a More Caring Society. In *Race, Class, and Gender*, 3rd ed., ed. P. S. Rothenberg, 461–475. New York: St. Martin's Press.

Sierra, Christine M. 1984. Chicano Political Development: Historical Considerations. In *Chicano Studies: A Multidisciplinary Approach*, ed. E. Garcia, 79–98. New York: Teachers' College Press, Columbia University.

Sierra, Christine Marie, and Adaljiza Sosa-Riddell. 1994. Chicanas as Political Actors: Rare Literature, Complex Practice. *The National Political Science Review* 4:297–317.

Simmons, Marc. 1979a. History of Pueblo-Spanish Relations to 1821. In *Handbook of North American Indians: Southwest* vol.9, ed. A. Ortiz, 178–193. Washington, D.C.: Smithsonian Institution.

———. 1979b. History of the Pueblos Since 1821. In *Handbook of North American Indians: Southwest* vol.9, ed. A. Ortiz, 206–223. Washington, D.C.: Smithsonian Institution.

Southwest Network for Environmental & Economic Justice. 1990. *Statement of Solidarity: People of Color Regional Activist Dialogue on Environment Justice*. Albuquerque, NM: Southwest Community Resources, Inc.: April 8.

SouthWest Organizing Project. 1995. Five Years Later: SWOP's Letter to the 'Group of Ten' Revisited. *Voces Unidas* 5(3): 1–4 Special Insert.

———. 1994a. The Selling of New Mexico. *Voces Unidas* 4(2): Special Insert (4).

———. 1994b. Grassroots Democracy in Action: An Interview with Richard Moore. *Voces Unidas* 4(3): 9 (3).

———. 1991. Carletta Tilousi: Fighting For Her People. *Voces Unidas* 1(4): 13.

Stack, Carol B. 1994. Different Voices, Different Visions: Gender, Culture, and Moral Reasoning. In *Women of Color in U.S. Society*, ed. M. Baca Zinn and B. Thornton Dill, 291–301. Philadelphia: Temple University Press.

Stanley, Liz, and Sue Wise. 1990. Method, Methodology and Epistemology in Feminist Research Processes. In *Feminist Praxis: Research, Theory and Epistemology in Feminist Sociology*, ed. L. Stanley, 20–59. New York: Routledge.

Steckenrider, Janie, and Seth Thompson. 1995. Gendered Agendas: An Exploration in the Social Construction of Political Life. Paper presented at the annual meetings of the Western Political Science Association in Portland, OR, March 16–18.

Stern, Paul C., Thomas Dietz, and Linda Kalof. 1993. Value Orientations, Gender, and Environmental Concern. *Environment and Behavior* 25(3): 322–348.

Steurnagel, Gertrude A. 1987. Reflections on Women and Political Participation. *Women & Politics* 7(4): 3–13.

Stiffarm, Lenore A. 1992. The Demography of Native North America: A Question of American Indian Survival. In *The State of Native America: Genocide, Colonization, and Resistance*, ed. M. A. Jaimes, 23–53. Boston: South End Press.

Strauss, Anselm, and Juliet Corbin. 1990. *Basics of Qualitative Research: Grounded Theory Procedures and Techniques*. London: Sage Publications.

Susser, Ida. 1988. Working-Class Women, Social Protest, and Changing Ideologies. In *Women and the Politics of Empowerment,* ed. A. Bookman and S. Morgen, 257–271. Philadelphia: Temple University Press.

Takakai, Ronald. 1990. *Iron Cages: Race and Culture in 19th-Century America*. New York: Oxford University Press.

———. 1987. Reflections on Racial Patterns in America. In *From Different Shores: Perspectives on Race and Ethnicity in America*, ed. R. Takaki, 26–37. New York: Oxford University Press.

Taliman, Valerie. 1991a. Greenpeace Officials Visit NM Victims of Economic Blackmail and Environmental Racism. *Voces Unidas* 1(3): 5 (3).

———. 1991b. Uranium Miners Made Profits, Left Town, and Stuck Natives with Open Pits and Contamination Worries. *Voces Unidas* 1(3): 6.

———. 1991c. Native American Activists and Supporters Protect Mother Earth. *Voces Unidas* 1(3): 9 (4).

Tarrow, Sidney. 1991. Understanding Political Change in Eastern Europe. *PS* 24(1): 12–20.

Taylor, Dorceta. 1993. Environmentalism and the Politics of Inclusion. In *Confronting Environmental Racism: Voices from the Grassroots*, ed. R.D. Bullard, 53–61. Boston: South End Press.

———. 1990. Can the Environmental Movement Attract and Maintain the Support of Minorities? In *The Proceedings of the Michigan Conference on Race and the Incidence of Environmental Hazards*, ed. B. Bryant and P. Mohai, 28–53. Ann Arbor: University of Michigan, School of Natural Resources.

———. 1989. Blacks and the Environment: Toward an Explanantion of the Concern and Action Gap Between Blacks and Whites. *Environment and Behavior* 21(2): 175–205.

Taylor, Lynda, et al. 1988. The Importance of Cross-Cultural Communication Between Environmentalists and Land-Based People. *The Workbook* 13(3): 90–93.

Tesch, Renata. 1990. *Qualitative Research: Analysis Types and Software Tools.* New York: Falmer Press.

Tesh, Sylvia N., and Bruce A. Williams. 1996. Identity Politics, Disinterested Politics, and Environmental Justice. *Polity* 18(3): 285–305. The Panos Institute. 1990. *We Speak for Ourselves: Social Justice, Race and Environment.* Washington, D.C.

Thomas, Sue. 1997. Why Gender Matters: The Perceptions of Women Officeholders. *Women & Politics* 17(1): 27–53.

———. 1994. *How Women Legislate.* New York: Oxford University Press.

Thomas-Slayter, Barbara, Esther Wangari, and Dianne Rocheleau. 1996. Feminist Political Ecology: Crosscutting Themes, Theoretical Insights, Policy Implications. In *Feminist Political Ecology: Global Issues and Local Experiences*, ed. D. Rocheleau, B. Thomas-Slayter, and E. Wangari, 287–307. New York: Routledge.

Thornton Dill, Bonnie. 1987. Race, Class, and Gender: Prospects for an All-Inclusive Sisterhood. In *From Different Shores: Perspectives on Race and Ethnicity in America*, ed. R. Takaki, 204–213. New York: Oxford University Press.

Tilly, Louise A., and Patricia Gurin. 1990. Women, Politics and Change. In *Women, Politics and Change*, ed. L. A. Tilly and P. Gurin, 3–34. New York: Russell Sage Foundation.

Tolleson Rinehart, Sue. 1992. *Gender Consciousness and Politics.* New York: Routledge.

Tong, Rosemarie. 1989. *Feminist Thought: A Comprehensive Introduction.* Boulder: Westview Press.

Trask, Haunani-Kay. 1993. *From a Native Daughter: Colonialism and Sovereignty in Hawaii.* Monroe: Common Courage Press.

Tronto, Joan. 1993. *Moral Boundaries: A Political Argument for an Ethic of Care.* New York: Routledge.

———. 1991. Changing Goals and Changing Strategies: Varieties of Political Activities. *Feminist Studies* 17:85–104.

United Church of Christ Commission for Racial Justice. 1987. *Toxic Wastes and Race in the United States: A National Report on the Racial and Socio-Economic Characteristics of Communities with Hazardous Waste Sites.* New York: United Church of Christ.

United States Department of Commerce, Bureau of Census. 1990a. *1990 Census of Population - General Population Characteristics, New Mexico.* 1990 CP-1–33.

United States Department of Commerce, Bureau of Census. 1990b. *1990 Census of Population - Social and Economic Characteristics, New Mexico.* 1990 CP-2–33.

United States Environmental Protection Agency, Office of Communications, Education, and Public Affairs. 1992. *EPA Journal* 18(1): (entire issue).

United States General Accounting Office. 1983. *Siting of Hazardous Waste Landfills and Their Correlation with Racial and Economic Status of Surrounding Communities.*

Van Liere, Kent, and Riley E. Dunlap. 1980. The Social Bases of Environmental Concern: A Review of Hypotheses, Explanations, and Empirical Evidence. *PublicOpinion Quarterly* 44(2): 181–197.

Waldowski, Paula. 1980–81. Alice Brown Davis: A Leader of Her People. *Chronicles of Oklahoma* 58(4): 455–463.

Wangari, Esther, Dianne Rocheleau, and Barbara Thomas-Slayter. 1996. Gendered Visions for Survival: Semi-arid Regions in Kenya. In *Feminist Political Ecology: Global Issues and Local Experiences*, ed. D. Rocheleau, B. Thomas-Slayter, and E. Wangari, 127–154. New York: Routledge.

Warren, Karen J. 1987. Feminism and Ecology: Making Connections. *Environmental Ethics* 9:3–20.

Wastl-Walter, Doris. 1996. Protecting the Environment Against State Policy in Austria: From Women's Participation in Protest to New Voices in Parliament. In *Feminist Political Ecology: Global Issues and Local Experiences*, ed. D. Rocheleau, B. Thomas-Slayter, and E. Wangari, 86–104. New York: Routledge.

West, Guida, and Rhoda Lois Blumberg. 1990. Reconstructing Social Protest From a Feminist Perspective. In *Women and Social Protest*, ed. G. West and R. L. Blumberg, 3–35. New York: Oxford University Press.

Westkott, Marcia. 1990. Feminist Criticism of the Social Sciences. In *Feminist Research Methods*, ed. J. McCarl Nielsen, 58–68. Boulder: University of Colorado.

Westra, Laura, and Peter S. Wenz. 1995. Introduction. In *Faces of EnvironmentalRacism: Confronting Issues of Global Justice*, ed. L. Westra and P. S. Wenz, xv-xxiii. Lanhman: Rowman and Littlefield Publishers.

Willard, William. 1984. Gertrude Bonnin and Indian Policy Reform, 1911–1938. In *Indian Leadership*, ed. W. Williams, 70–75. Manhattan: Sunflower University Press.

Wilson, Larry. 1989. Moving Toward a Movement. *Social Policy* 20(1): 53–57.

Witt, Linda, K. M. Paget, and G. Matthews. 1994. *Running as a Woman: Gender and Power in American Politics*. New York: Free Press.

Women's Environment and Development Organization. 1995. *Turn the Words into Action!: Highlights from the Beijing Declaration and Platform for Action.* New York: November 30.

Woodbury, Richard B. 1979. Prehistory: Introduction. In *Handbook of North American Indians: Southwest* vol.9, ed. A. Ortiz, 22–30. Washington, D.C.: Smithsonian Institution.

Young, Kate. 1988. Reflections on Meeting Women's Needs. In *Women and Economic Development: Local, Regional and National Planning Strategies*, ed. K. Young, 1–31. New York: Berg/UNESCO.

Zavella, Patricia. 1988. The Politics of Race and Gender: Organizing Chicana Cannery Workers in Northern California. In *Women and the Politics of Empowerment*, ed. A. Bookman and S. Morgen, 202–224. Philadelphia: Temple University Press.

Zeitler, Amy. 1996. Representation in the Era of Expertise: Citizen Input and Environmental Policy Making. Paper presented at the annual meetings of the Western Political Science Association in San Francisco, CA, March 14–16.

Zimmerman, Michael E. 1987. Feminism, Deep Ecology, and Environmental Ethics. *Environmental Ethics* 9:21–44.

WEBSITES ACCESSED

http://www.doi.gov/bia/stats930.html
http://www.fw.umn.edu/indigenous/gloss.html
http://www.msnbc.com/onair/nbc/nightlynews/fleecing/newdeal.asp

Index

abolition 24
aboriginal 14, 17
  languages, use of 18
  women 15–16
abortion 25
  Freedom of Access to Clinic Entrances Bill 32
  National Abortion Rights Action League (NARAL) 26
  rights 119
acculturation 11
acequias (irrigation canals) 148
active interview 49
acts
American Indian Religious Freedom Act (*see also* Native American; traditional religious practices) 163
  Clean Air Act 142
  Clean Water Act 142
  Elective Franchise Act 20
  Endangered Species Act 146
  Family and Medical Leave Act, the 32
  Freedom of Access to Clinic Entrances Bill (*see also* abortion) 32
  Indian Child Welfare Act 162
  Indian Education Act of 1972 (*see also* Native American; education) 129
  Indian Mineral Development Act 163
  Indian Reorganization Act (IRA) (*see also* Native American) 20–21
  Violence Against Women Act, the 32
  Women's Health Equity Act 32
agropastoral practices, traditional 98–99
American Indian Movement (AIM) 86, 87, 90
American Indian Religious Freedom Act. *See also* Native American; traditional religious practices 163
associations 20, 25, 34, 44, 66, 68, 89, 108, 176

barrio 33, 58, 65, 95, 101
barter economy 18
Beijing Declaration and Platform for Action 188
bilingual 83, 164
biodiversity, maintaining 142, 146
Black Power movement 24, 26, 91
boarding schools. *See also* Native

American; education 18, 21
Browner, Carol. *See also* Environmental Protection Agency 62
Bureau
Census, of the 5, 46, 84
Indian Affairs, of 21, 23, 122
Land Management, of 150

campaigns, for public office 30, 55, 77, 81, 92, 102, 108
Catholicism, conversion to 14–15, 17
Center for the American Woman and Politics (CAWP). *See also* Rutgers University 31
characteristics, demographic of participants 52, 71–72, 87, 108
Chicana 46, 54, 112–113, 117, 120, 129, 139–140
Chicano movement 25, 86, 90–91, 128
child care 25, 81, 118, 177
funding 119
children 15, 18, 22, 24, 27, 31, 55, 73, 78, 80–84, 94, 97, 101, 107–108, 118–119, 121, 123–126, 128–129, 132, 142, 157, 161–164, 169, 177
children's rights 102, 161
church 86, 89
citizen involvement 3, 19, 162, 168, 177
Citizen's Clearinghouse for Hazardous Wastes 166
citizenship 23, 35
dual citizenship 10
teaching of 48
civic groups 20, 86, 89, 93, 171
civic obligation/responsibility 8, 22, 85, 93–95, 104, 109, 182
civil liberties 22
civil rights 20, 22–24, 27, 33, 86–87, 90, 127, 150, 172, 177
class 24–30, 34, 37, 39–41, 54, 87, 100–101, 106, 108, 111–113, 126, 129–130, 138–143, 145, 148, 155–156, 175, 177, 184–185
Clean Air Act 142
Clean Water Act 142
effluent discharges
groundwater contamination 7, 168
pollution 6–7, 24, 34, 143, 168
quality standards 34, 170
Clinton Administration 30
coalitions 6, 30, 171, 176
collective action 28–29, 117, 152–153, 155
colonial hierarchy 16, 22, 116
colonization 15, 17, 33, 35, 184
legacy of 10, 18, 20, 130
pobladores (settlers)16–17
communities 5–9, 12, 14–17, 20, 23, 26, 33–35, 39, 45, 71, 87–88, 93, 95–96, 105, 107,-109, 112, 115, 117, 123, 126–127, 129, 133, 138–139, 141–145, 147–148, 151, 156, 159–163, 168, 178, 182, 185–189
economically depressed 98
of color 3, 6, 43, 107, 145, 148, 150, 154, 163–164, 166, 168, 170, 191
community 7, 16, 20, 23–24, 26, 28, 30, 35–37, 48–49, 58, 60–61, 63, 67, 69, 72, 76–77, 82, 85–89, 92–93, 95–97, 99–100, 102, 104–107, 115–117, 120–122, 124, 127, 129, 137–140, 151, 156–157, 159, 165–166, 169–171, 174, 176–177, 182–184, 186
development 64, 160–161, 187
health/safety 48, 65, 101
Latino 115
leadership 5–6, 65, 67, 82
members/residents 3, 27, 61, 90, 99, 114, 157, 159, 163–164, 169, 171–172, 176–178, 183, 185, 187–188
organizations 30, 40, 44, 47, 90, 176
services 33, 191

work 9, 27, 68, 94, 116–117, 125, 153–154, 183
community-based activism. *See* community work
comparable case, sample approach 30, 43
compensation 60
toxic victim 100
Congress 7, 20, 34, 87, 129
United States 103rd 32
Congresswomen's policy agendas
evidentiary rules in the criminal trials of sex offenders 32
Family and Medical Leave Act, 32
Freedom of Access to (abortion) Clinic Entrances Bill (*see also* abortion) 32
Individual Retirement Accounts for homemakers 32
insurance coverage of mammograms 32
North American Free Trade Agreement (NAFTA) 32
sexual harassment of women in the military 32
Violence Against Women Act, 32
Women's Health Equity Act 32
conquistadores 14–15, 112
consensus building 21, 34
conservation 21, 57, 146–148, 160, 166–167, 187
sustainable use 47, 144–145, 147–149
conservative 11, 57, 132, 149, 151
constituency 11, 65, 104, 149, 152, 168–169, 171
constitution 19–20
New Mexico state framers 18
tribal 21, 163
contamination 66, 145, 166
groundwater 7, 64, 168
soil 64
corn 14, 128
Council of American Indians. *See also* Native American 87
countersocialization 29
crime prevention 33, 164
cultural preservation 28, 33, 48, 64, 69, 94, 98–99, 115, 128,144, 150, 160–164
cultural/religious artifacts. *See also* Native American 48, 98, 150, 162, 171–172
cultural traditions 14, 21, 69, 117, 147–148, 161–162
curanderas/os (healers) 148

Davis, Alice Brown 34
day schools. *See also* Native American; education 21
deductive comparative methods, theory. *See also* theory 42, 50
democracy 3, 67, 105, 107
environmental 177
participatory 9, 35, 169, 178
representative 5, 11, 21, 31
democratic 34, 38, 61–62, 96, 105, 158, 162, 165, 178, 189, 191
process 39, 152
structures 69, 163
party, political 82, 91, 103
developers
private 64
development 6, 18, 21, 29, 35, 38, 64–66, 91, 95, 98–99, 105, 108, 121, 130, 139, 146, 149–150, 152, 158, 160–161, 164,168, 174, 177–178
urban 150, 168
urbanization 57
disabled, rights 23
disconfirming cases 43
discrimination 34, 53, 59, 79, 111, 120, 134–135, 140
discriminatory zoning 60
disproportionate risk. *See also* environmental racism 59–60
dissent 107
distribution of wealth, resources, influence and power 21, 39, 62, 149
domestic violence 25, 48, 131, 162
dominant political culture 5, 10–11, 53

double minorities 5
dual citizenship 10
dumps, radioactive 99

economic(s) 5, 14, 17–18, 20, 23, 28–29, 32–34, 57, 60, 64, 99, 107, 111–113, 116, 128, 130, 138, 142–144, 147–150,152, 154, 159, 163, 165–166, 169, 177
  conditions 3, 9–10, 64, 94,129, 154, 159, 178, 187
  constraints, on women 25, 121
  dependence, of New Mexico 64
  development 21, 35, 65, 99, 121, 128, 147, 164–166, 168, 187
  equity 10, 26–27, 60, 141, 146, 149, 154, 178
  factors 153
  growth 130, 145
  justice 65–66, 148, 188
  security 33
  elderly women, for 119
economically depressed communities 66, 98
economies, local 17–18, 146
education 21–22, 25, 33–35, 54, 58, 60, 108, 115, 123, 129–130, 153, 161, 163–165, 169,
  level of 34, 73–74, 76, 78, 87, 106, 158, 184
  programs
    bilingual for Navajo children 164
    effectiveness, of grassroots/ public officials to achieve change 52, 173, 191
egalitarianism 102
elder, role of 54
elected officials 3–5, 8, 32, 40, 44, 48–49, 88–89, 102–104, 116, 123,156, 160, 162–164, 168, 171–173, 176–177
elected office/positions 20, 23, 26, 30, 32, 45, 47, 158, 176, 182, 185, 190–191
elections. *See also* suffrage; voting 22, 36, 48
  federal 19
  municipal 91
  school board 20
employment 6, 25, 89, 115, 140, 159
equal employment opportunities 119, 177
opportunities 64, 161
empowerment 33, 47–48, 59, 95, 96, 98, 102, 119, 139, 144, 160–163, 177
Endangered Species Act 146
energy sources, production of renewable 166–167
enslavement of Indians 14, 17
environment 16, 20, 23, 30, 48, 55, 57–58,60, 62–63, 66, 77, 94, 101, 116, 130, 141–142, 145–146, 148–151, 155, 157–159, 164, 167, 170, 181, 188
  definition of 144
environmental 7, 28, 30, 37–38, 43, 44, 47–49, 57–61, 64, 66, 68, 76, 107–108, 142, 144–149, 151, 155, 160, 165–166, 168, 170, 176–177
  conditions/quality 3, 9, 27, 64, 159, 166, 178, 183
  degradation 6, 64, 101, 151, 154–155, 181
  health 48, 99, 165, 187
  ideologies, of the leaders 101, 145–150, 187
  issues 6–7, 58, 61, 67, 91, 124, 143, 145, 159, 167, 181
  justice movement 3–4, 6–7, 24, 27, 30, 33, 46, 56–59, 61–69, 71, 90, 97, 99, 107–108, 143–144, 148–150, 153–155, 157, 159, 161, 166–169, 171, 173, 175, 177, 179, 182, 187–188, 191
  policymaking 4, 6, 27, 30, 39, 43, 59, 168, 177, 181
  policy/politics 6–8, 27, 43, 56, 72,

108, 148, 151–152, 156, 159, 161, 163, 165–167, 177,187
protection 33, 47–48, 58, 62, 144, 155, 166–167
general/environmental standards 7
state 171
quality 5, 49, 101, 128, 142, 148, 160, 165–167, 187
racism 59, 62, 145, 149, 185
disproportionate risk 59–60
threats 59, 63
Environmental Protection Agency 59, 62, 170
environmentalism 11, 13, 52, 112, 141–142, 144–145, 148,186–187
mainstream 61, 142, 146
environmentalists 8, 11, 57–58, 141–143, 146–148, 171, 187
indigenous 143
third world 143–145, 149, 187
ethnicity 4, 16, 27, 30, 32, 35, 37, 41, 44, 46, 52–54, 58–59, 61, 85, 88, 100, 111–114, 117, 125, 154, 181–187
Euro-American women 26, 46, 53, 94
executive
agencies 107
branch 49, 163
order 63
executives 158, 177

family 8, 15–16, 23, 28, 31, 34, 55, 67, 76–79, 80–83, 85–86, 88, 92, 97, 109, 113, 124–125, 127–128, 131, 135, 157, 162, 177, 183–184, 186
extended 79, 127
make-up 80
obligations 15, 77, 83
planning 24
faith 17, 97
federal 21, 33–35, 59, 62–64, 129, 147–148, 163, 170
Advisory Committee on Environmental Justice 66
agencies 34, 49, 63
elections (*see also* suffrage and voting) 19
jurisdiction 20
policies 189
statutes 155
female consciousness 126
feminism 11, 13, 25–26, 29, 52, 112, 124–125, 127–132, 134–136, 138–141, 150, 186–187
alternative 29, 187
ambivalence towards 131, 133–134
Chicana 129
insurgent 139–140
liberal 139–140
Western 140
white middle-class 130, 133, 138, 141–142
feminist 6, 11, 25, 29, 35, 41, 50, 96, 111, 113, 115, 120–121, 127–141, 158, 186–187
closet 135
maternal 124–125
theory 4, 9
radical 127
scholars/theorists 4, 9–10, 26, 29, 36–37, 39–40, 55, 68, 77
feminists 8, 11, 120, 186–187
indigenous 132, 138–140, 187
First National People of Color Environmental Leadership Summit (1991) 66
Franciscan missions 15
Franciscan missionaries 16

gay and lesbian rights 23–24
gender 4, 10, 14, 27, 29–31, 35–37, 39–40, 50, 53–54, 72, 78, 88, 100–101, 106, 109, 112, 118–119, 121, 129–130, 135, 141, 154–156, 158, 177, 181, 186
consciousness 29, 113, 118–119, 124, 126–127, 132–133, 135, 141, 150, 186
Rinehart, Sue Tolleson 113, 126

differences 157
equity 27, 120, 128, 146, 148, 154, 169, 185, 187–188
gap 30
identity 8, 10–11, 13, 37, 52, 85, 102, 112, 117–118, 120–121, 178, 181–182, 186
interests 126
practical 127
strategic 127
politics 4, 189
subordination 111
general population 46, 72
generalization, of findings 43
generations, future 66, 98, 155, 157
genocide 33, 66
Gibbs, Lois 165
grassroots 7, 21, 26–27, 30, 36, 44–45, 59, 66–67, 79, 90, 94, 104–107, 151, 155, 176, 178, 182
activists 4–5, 7–13, 40, 44, 47, 69, 83, 89–90, 104, 107, 156, 160, 162–165, 168–171, 173–177, 182–183, 185, 187–189, 191
groups/organizations 6, 27, 29, 32, 39, 44–45, 47–48, 60, 63–68, 89, 107, 172, 176, 182, 188, 191
mobilization 23, 63
organizers 3, 13, 105
politics 4, 9, 26, 30, 35, 40, 56, 67, 69, 78. 182–184
grazing, livestock 15
Great Spirit. *See also* Native American; traditional religious practices 129
Green party. *See also* party 182
grounded theory. *See also* theory 40–43, 50
groundwater contamination. *See also* Clean Water Act 7, 168
group 4, 9, 11–12, 24, 28–29, 31–32, 45, 49, 54, 65, 69, 73–74, 76, 80, 83, 93–94, 96–97, 103, 108–109, 151–155, 157–158, 160–162, 171–172, 174–176, 183, 185–186
consciousness 152
identity 153
norms 153
groups 39–40, 44–48, 54, 58–61, 63–65, 67–68, 74–75, 78, 81, 86, 89–91, 94, 96, 107–108, 151–156, 160–163, 165, 168–169, 171–172, 174–178, 181–184, 186, 189, 191
marginalized 35–36, 50, 97, 107, 152, 159, 163, 178
service 89–90, 92
unrepresented 152, 158

haciendas 15–16
hazardous waste 7, 60–61, 64, 166
health care 33, 35, 164–165
Hill, Anita 30
Hispanas 4–6, 8, 10, 23, 25, 30, 32–35, 37, 39, 44, 46, 56, 71, 81, 86, 89, 91, 94–95, 97–99, 102–103, 163, 165, 167, 178, 183–187, 189, 191
Hispanic women 3–5, 7–9, 11–13, 16, 20, 23, 25–27, 32–35, 37–38, 44–46, 69, 81, 87, 181–182, 184–185, 187–191
double minorities 5
Latinas 35
homemaker, role of 67, 108
housing 22, 35, 60, 164–165
low-income 64
human rights 160–164, 167–168, 188

ideological orientations, of participants 52
ideology 4, 7–8, 25, 29, 58, 61, 66, 69, 78, 155, 181–182, 186–187
implementation 167–168, 177–178
of policy directives 39
incentives 152–153, 155–156, 178
income 34, 58–59, 63–64, 76, 87, 159, 168–169
household income 74–75, 87

median 74
Independents, voting affiliation. *See also* elections; party 103–104, 107
Indian Child Welfare Act 162
Indian Education Act of 1972. *See also* Native American; education 129
Indian Health Service 129
Indian Mineral Development Act 163
Indian pueblo(s) 6–7, 13–14, 16, 22, 34, 454, 81, 84, 87, 96–98, 101, 106, 108, 113, 115, 118, 121–124, 132, 148, 170,
Indian Reorganization Act (IRA) 20–21
Indian reservation(s) 18, 20–21, 84, 87, 99, 103, 115, 121, 123, 133, 142–147, 149–150, 155, 163, 170
Indian sovereignty 33, 64, 107–108, 114, 121, 128–129, 150, 155, 162–163, 185
Indigenous 4–12, 14, 18, 21, 27, 30, 33–35, 39, 44–47, 56, 69, 71–72, 83, 90–91, 113–115, 117–121, 123, 127–128, 130–133, 138–140, 142–144, 146–149, 155, 157, 161–163, 165, 167, 174, 182–186, 188–189
indigenous peoples 21, 48, 64–65, 91, 112, 144, 146, 163, 166–167
  commercialization of 64, 98
  environmental rights 47–48
inductive strategy, theory. *See also* theory 42, 50–51
industrial 69, 165
  air emissions 64
revolution 57
industries 60, 166, 181
  high tech
  polluting 59, 101
influence 4–5, 8–10, 12, 14, 17–18, 20–21, 25, 29–30, 32, 37, 39, 59, 61, 63–64, 69, 71, 78–79, 87–89, 92, 96, 102, 107–109, 112, 122, 132, 152, 158–159, 165, 169, 171–172, 174, 176–177, 182–184, 186, 189, 191
inhabitants, original of New Mexico 13–14, 17
institutionalized politics 12, 40, 107
interview guides 51
intimacy/self-revelation, in the interview 54–55
investigative process 42

job 72, 76,81, 83, 100, 117, 132, 134–135, 186
  blackmail 60
  training 25, 118–119
judicial branch, of government 49, 163
Juris Doctor degree 74

Kirtland Air Force Base 7
kiva 13

La Raza Unida party. *See also* party 86, 90
labor 15–17, 20, 24, 83, 86, 90, 99, 127, 140, 171
  movement 24, 33
  party. *See also* party 103–104
LaDuke, Winona. *See also* Native American 144
land 6, 18, 61, 64, 66, 91, 99–100, 133, 144–145, 147–150
  grants (*mercedes*) 15, 64, 147–148, 161–162,
  ownership of, 16
  use 33, 58, 60, 145, 166, 167–168, 181
    planning 65, 159
  water/mineral rights 64, 91, 115
    return of/to indigenous tribes 48, 171–172
language(s) 18–19, 34, 52, 83–84, 99, 112, 117, 121, 162, 164, 185
  English 19, 34, 83–84
  North American 83
  policies 91

Spanish 19, 46, 83, 164
use of aboriginal 18, 83–84, 114
Latinas 35
leaders 3–5, 7–13, 22–24, 26–27, 29–30, 32, 34–35, 38–39, 43–46, 49, 51–53, 55–58, 66, 69, 71, 78–79, 87–88, 96–98, 105, 121, 137, 157, 159, 172, 174, 177, 181–182, 184–185, 190–191
community 7, 37, 67, 88
New Mexico 9, 11, 18, 32, 44–48, 69, 72–86, 88–99, 101–109, 112–120, 122–124, 127, 130–150, 154–157, 159–163, 165–169, 171–176, 178, 182–191
leadership 4–7, 9, 11, 18, 21, 26–27, 30–35, 39, 45, 47, 56, 65–69, 71–73, 76, 78–79, 85, 87–88, 90–91, 95–97, 101, 104–109, 122, 124–125, 133, 155, 157, 169–170, 172, 177–178, 181, 183–185, 187, 189, 191
community 5–6
experience 12, 102, 107, 182
opportunities for 44, 68, 92, 108
positions 134, 138–139, 177, 182, 189, 191
roles 3, 34–35, 58, 69, 85, 95, 97
skills/confidence 48, 182
legal rights 17–18, 128
legislation 21, 26, 32, 61, 118, 134, 137, 142, 158, 162–163, 168, 171, 176, 186, 190
legislative branch 49, 98, 107, 163, 190
debates 178
politics 137
sessions 135
legislators 31, 100, 134–135, 137, 158, 170, 176
legislature 31, 116, 135, 173
state, New Mexico 134, 190
volunteer, in New Mexico 156
liberal 11, 57, 139–140, 158
life experience, of participants of study 49
local
economies 17–18, 99, 146
elections 20
expertise/knowledge 30, 58
officials 108, 171
politics 3, 6, 38, 63, 86, 91, 184
school board 48, 83, 108
member 106
locally undesirable land uses (LULUs) 6, 61
Logic of Collective Action, the 152

male 14, 30–31, 35–36, 40, 58, 69, 78–79, 88, 104, 116, 118, 120, 122–123, 126, 128, 130–131, 133–137, 139, 158, 190
activists 9
masculinity 16
mass movements 111
media, the 30, 54, 63, 108, 151, 166, 168, 170, 173
Medicine Men. *See also* Native American; traditional religious practices 86
men 14–16, 19, 22, 26–27, 29, 31, 37, 40–41, 53, 68, 78–79, 82, 86, 88, 96–97, 112, 115, 118, 120–121, 123–125, 127–137, 139, 157–159, 189–191
of color 139
menopause 53
mentoring 77, 88–89, 138, 169, 173, 177
Mestiza 16, 46
Mexican 15, 17–18, 35, 111, 113, 146
-American 18, 112–113, 133, 159
-American War 17
-American women 33–34, 40, 46, 139
period (1821–1846) 17
military 32, 89, 91, 101, 145
mineral rights 48, 168
mining 6, 64, 99, 167, 181
minority women, politics of 28, 182

Mother Earth. *See also* Native American; traditional religious practices 124
motherhood 28, 33, 124–125, 128
mothers 24, 27, 68–69, 78, 83, 86–87, 118, 120, 123, 126, 128, 138, 159–160, 164
  single 82–83, 118, 162
motivation 68–69, 76, 98, 113–114, 118–119, 153, 156
movement(s) 25, 30, 58, 62, 66, 78, 86, 90–91, 129, 131–133, 145, 166, 187, 192
  American Indian 24–26, 86–87, 90, 128
  Black Power 24, 26, 91
  Chicano 24, 26, 86, 90–91, 128
  Civil Rights 27, 61, 86, 90, 172
  environmental justice 4, 6–7, 24, 27, 30, 56–59, 61–69, 99, 143–145, 148–150, 154–155, 159, 166–168, 182, 188, 191
  farm workers 24
  labor 24, 33
  mass 111
  nationalist movements of the 1960s/70s 128
  native rights 53
  peace 90
  social 4, 9–12, 23, 25, 29, 36, 59, 61–62, 67, 86, 89–92,108, 129, 152, 154–155, 182–183, 187
  Third World 91
  Women's 9, 24–28, 37, 62, 91, 128–130, 133, 136, 138–140, 158
Ms. Foundation 133
multilingual 83–84
municipal waste facilities 60

national organizations
  Citizen's Clearinghouse for Hazardous Wastes 166
  National Abortion Rights Action League (NARAL) (*see also* abortion) 26
  National Organization of Women (NOW) 26
  North American Indian Women's Association 129
  Ms. Foundation 133
  Sierra Club 66, 172
nationalist movements of the 1960s/70s 128
Native
  -American legal status 10
  -American political status 10, 120
    lands 33, 98–99, 161–162
  rights movement 53
  social structures 16
Native America(n)
  All Indian Pueblo Council 123
  American Indian Movement (AIM) 87, 90
  American Indian Religious Freedom Act (*see also* Native American; traditional religious practices) 163
  American Indian 3–7, 9,-10, 12–13, 19, 21, 24–26, 32–33, 37–40, 45–46, 53, 71, 87, 90, 113, 115, 120, 127–130, 133, 145, 150, 163, 167, 178, 181–183, 187–191
  Bureau of Indian Affairs 21, 23, 122
  Clan Mothers 86
  constitution, tribal 21, 163
  Council of American Indians 87
  Cultural/religious artifacts 48, 98, 161–162, 171–172
  double minorities 5
  dual citizenship 10
  education 21–22, 33–35, 73, 123, 129
    boarding schools 18, 21
    day schools (*see also* education) 21
    Indian Education Act of 1972 (*see also* education) 129
  egalitarian tribes 21, 123, 149
  elder, role of 54

enslavement of Indians 14, 17
Great Spirit (*see also* traditional religious practices) 129
Indian Child Welfare Act 162
Indian Health Service 129
Indian sovereignty 33, 64, 107–108, 114, 121, 128–129, 150, 155, 161–163, 185
inhabitants, original of New Mexico 13–14, 17
Tanoans (Tiwas, Tewas and Towas) 13
Keresan 13
Zunis 13
kiva 13
LaDuke, Winona 144
Medicine Men (*see also* traditional religious practices) 86
Mother Earth (*see also* traditional religious practices) 124
North American Indian Women's Association 87, 129
pueblo(s) 6–7, 13–14, 16, 22, 34, 454, 81, 84, 87, 96–98, 101, 106, 108, 113, 115, 118, 121–124, 132, 148, 170,
racial/ethnic identity 8, 10–11, 13, 44, 46, 83, 85, 109, 112–115, 118, 120–121, 178, 181, 185–186
reservation(s) 18, 20–21, 84, 87, 99, 103, 115, 121, 123, 133, 142–147, 149–150, 155, 163, 170
Spirit Mother 128
Stewart, Irene 22
Tribes
Athabascans 13–14
Apaches 13, 16
Navajos 13, 16
Bad River Chippewa 45
Blackfoot 45
Comanche 16, 45, 84, 113, 117, 123, 127
Diné (Navajo Nation) 6–7, 21–23, 45, 84–85, 92, 164
Kiowa 45, 84
Mescalero Apache 7
Mohawk 84, 115, 123
Western Shoshone 45, 123
traditional religious practices 18
Wauneka, Annie Dodge 22
natural resources 14, 33–34, 37, 48, 57, 101, 146–147, 149, 160, 166–167, 187
nature 16, 21, 31, 35, 50, 56, 99, 103–104, 127, 140, 143, 145–146, 149, 152, 169, 182
neighborhood 47, 64–65, 67, 77, 89, 159
activists 4
associations/groups 7, 65, 89, 108, 176
safety 48, 64, 67, 115
networking 170, 173
New Mexico
Environment Department 170
Girls' State 88
leaders 9, 11, 18, 32, 45–46, 69, 72–86, 88–99, 101–109, 112–120, 122–124, 127, 130–150, 154–157, 159–163, 165–169, 171–176, 178, 182–191
politics 5, 7, 13, 43, 45, 64, 69, 72, 179, 184, 191
population 5, 39, 45–46, 72–75, 77, 83–84, 173
Nineteenth Amendment 20
non-profit organizations 47, 76, 89, 156
nuclear 64, 187
power plants 159
weapons 6, 33, 64
nurturer, role of 19, 69

office 19, 31–32, 35, 60, 62, 66, 88, 121, 124, 137, 158, 164, 173, 188, 190
appointed/elected 30, 45, 86, 176, 185, 190
Environmental Equity, of 62

political 22, 34, 88
public 8, 20, 22, 26, 30–31, 34–35, 48, 73, 79, 88, 102, 104, 108, 134–135, 158, 169, 172–173, 182, 184, 191
tribal 22
opportunistic selection, sample approach 45
oppression 29, 111–113, 116, 122, 125, 127, 129–130, 139–141, 154
oral history 41, 50, 56
outsiders 11, 137, 169, 171, 173, 184

participation in decision making, public
partisanship 103
party,
Democrat 103
Young Democrats 86
Elites 107
Green 27, 37, 104, 182
Labor 104
La Raza Unida 86, 90
Republicans 103
party elites 107
pastoral cultures 99
pastores (sheep growers) 147
patriarchal
institutions 14
values 16
Peace movement 24, 28, 86, 90
People of Color Regional Activist Dialogue 66
personal
growth 153
narratives 50, 56
Petroglyph Park 150
Pluralism 5
policies, public 24
child care 25
divorce 25
education 25
employment 25
insurance 25
job training (*see also* employment) 25
welfare 25
policy 10, 11, 12
agendas 29–30, 32, 44, 65, 112, 135, 150, 156–157, 159–161, 164,166, 168–169, 175–176, 179, 181–182, 186–189, 192
forums 182preferences of American Indian women and Hispanas 4
priorities 7, 31, 38, 124
recommendations 66, 156
political
activism 4, 5, 10, 25, 27, 34
analysis 41, 56
consciousness 29, 68, 104–106, 111, 113, 114, 118, 119, 124, 126, 127, 132, 133, 135, 144, 150, 177, 186
disenfranchisement 22, 154
efficacy 67, 153, 185
empowerment 48, 53, 56, 59, 95–96, 98–99, 102, 119, 139, 144, 160, 161–163
expression
demonstrations 55–56, 65, 87, 170, 172
petitions 65
pickets 65
goals 5, 8–13, 27–29, 31, 38, 45, 52, 56, 60–62, 65, 78, 107, 108, 112, 130, 133, 135, 138, 143, 146, 148, 150, 158, 160–161, 163, 168, 169, 174, 176, 182, 186–188
involvement
history of Indians
history of Hispanas
history of participants
resources 14–16, 18, 21, 28–29, 33–35, 37, 39, 48, 56–57, 62, 67, 91, 96, 98–101, 107, 114, 141–144, 146, 147, 149, 153, 160, 166, 167, 169, 171, 176, 177, 187, 191
science, discipline of 4, 26, 35, 36
socialization 4, 8, 29, 52, 56, 69, 72, 85, 86, 106, 108, 111, 124,

136, 181, 183, 189
strategies 5
conventional (or formal) 172, 172
unconventional (or informal) 13, 49, 52, 171–172, 175
strategy 15, 31, 42, 43, 45, 52, 117, 156, 171, 174, 175, 176
theory 4, 5, 9, 27, 35, 38, 40, 41, 42, 43, 50, 51, 127, 152–154, 178, 189
trajectories 13, 30, 32, 69, 72, 79, 85, 108, 181, 182, 184, 191
politics
electoral 4–5, 8, 12, 19, 20, 23, 26, 30–31, 34–35, 38, 44–45, 56, 69, 73, 78, 89–92, 102, 104, 137, 149, 169, 172, 174–176, 178, 180, 182–184, 188–190
institutionalized 107
polluters 35, 57, 62, 176
poor 6, 26, 57, 59, 60, 61, 63, 66, 87, 97, 101, 143, 150, 152, 159, 163, 159
population 5, 17,18, 19, 39, 43, 45, 46, 60, 63, 72, 118, 158, 181, 184, 190, 191 poverty 60, 62, 148, 154, 159, 163, 165
power 5, 6, 14, 16, 18, 19, 20, 21, 24, 26, 27, 29, 38, 39, 40, 48, 54, 56, 67, 78, 87, 96, 103, 104, 105, 106, 107, 112, 126, 130, 131, 134, 149, 153, 157, 159, 163, 167, 177, 178, 182–184, 186
prejudice, racial 135
preservation
religious/cultural 33, 48, 64
wilderness areas 57, 142, 144, 146
president
tribal
United States 30, 63
public
Citizen 3, 19, 38
Funds 168
health/safety 23, 47, 161, 168
hearings 166, 169–171, 173, 176
lands 147, 150
officials 4–5, 7–13, 29, 38, 40, 43–44, 48–49, 51, 56, 69, 72–77, 89–93, 95, 99, 103–105, 110, 114–115, 119–120, 132–133, 138, 142, 145–146, 148, 150, 162–165, 168–169, 171, 173–177, 180, 183, 185, 187–189, 192
policy 3, 5, 8, 11–12, 24–27, 30–32, 35, 52, 69, 71, 105, 107–108, 119, 121, 150 155, 160, 169, 171, 175–180, 183, 186–189, 192
sector 75, 89, 110
services 22, 161–163, 180

qualitative approaches, to theory. *See also* theory
quality of life 3, 8, 9, 12, 38, 108, 109, 117, 127, 141, 150, 156, 162, 164, 182, 186, 188

race 4, 19, 29, 32, 60, 61, 63, 72, 77, 106, 111, 112, 113, 125, 129, 130, 140, 141, 145, 154, 177, 182, 187, 189
racial/ethnic
consciousness, self-labeling 3, 113, 114
identity (*see also* Native Americans)
minorities 5, 6, 26, 27, 58, 59, 62, 139, 143, 152, 184, 185, 187, 189, 191
salience, self-labeling10, 11, 113, 114
racism 25, 26, 33, 59 62, 66, 77, 90, 112, 114, 115, 116, 117, 129, 130, 138, 139, 145, 149, 154, 185, 186
radiation exposure 64, 165
radical feminist theory. *See also* feminist 127
radioactivity 165
ranchers 144, 147, 150

rape 17, 53
rational choice theory. *See also* theory 152–154
RCRA permit 166
reduction, of waste 64
reforms
child labor 24
orphanage 24
mental institutional
prison 24
region 13, 17, 18, 30, 47, 60, 112, 113, 147
regulations
health and environmental 60
land use 58religion 16, 18, 19, 38, 112, 132, 162
religious groups 65, 89, 171
repatriation 161, 162
replication strategy 43, 45
Republicans. *See also* party 103
researcher, role of 4, 38
respondent, role of
rights
civil 57
gay and lesbian rights 23–24
Riley, William. *See also* Environmental Protection Agency 62
Rinehart, Sue Tolleson. *See also* gender consciousness 9, 36, 38, 113
Rio Grande 168
role(s)
domestic role of women 83
researcher 4, 38, 42
respondent 49
sex role 8, 10, 29, 37, 77
storyteller 54
women, of 5, 7, 9, 13,15, 18, 20, 21, 23, 24, 26, 27, 33, 34, 34, 35, 36, 83
role models
community leaders 7, 37, 88
fathers 88
female 7, 8, 9, 14, 22, 23, 30, 31, 32, 35, 38
friends 55, 102
grandparents 84, 88, 92
mothers 24, 27, 33, 38, 118
other family members 81, 85, 88
siblings 88
teachers 88
Rutgers University. *See also* Center for the American Woman and Politics (CAWP) 31safety and emergency responsiveness concerns

science 4, 26, 35,36, 36, 38, 40, 41, 99, 144, 145, 152, 175, 188, 189
secretary of state 48, 117
self-determination 35, 38, 48, 140 160, 161–162
self-interest, individual 154
self-labeling
race/ethnicity 27, 30, 35, 37, 38
racial/ethnic consciousness 10,11, 38
racial/ethnic salience
senior citizens 161
service groups 29, 90, 92
sex
egalitarian norms 14
female-male reciprocity 14
sexism 58, 66, 77, 112, 116, 127, 129, 130, 139, 154
sexual discrimination 53, 79, 134, 135
sexual orientation 29, 30, 112, 141,
sexuality 106, 112, 127
Sierra Club 66, 172
site clean-up 168
social
barriers 23, 28, 122
class 87, 108, 110, 112, 150, 184–185, 192
consciousness 3, 104
justice 5, 10–11, 24–26, 33, 38, 43–44, 47,49, 56, 65–67, 70, 97, 108, 110, 130, 148, 150, 178, 180, 185, 192
movement 12, 38, 61, 86, 89–90, 92, 110, 129, 150, 155, 182–183, 192

movement activists 12, 86
programs 110, 159, 161
psychological differences 125
reformers 24
sociopolitical situation 114, 118
soil erosion 64, 167
SouthWest Organizing Project (SWOP) 64, 65, 66, 71, 151
Southwest Network for Environmental and Economic Justice (SNEEJ) 65, 66
sovereign nations 121
Spain 14, 17, 38, 64
Spanish
conquest 14, 38, 121land grants (*mercedes*) 64, 147–148, 161–162, 180
law 17, 21, 38
spiritual
beliefs 18, 38, 101, 145, 149
Spirituality 90, 97, 98, 101, 127, 149, 150
status 10–11, 15–16, 25, 29–30, 32, 34, 38, 58–60, 67, 70, 80, 87, 96, 108–110, 118, 120–121, 123, 126, 129, 134–135, 137, 140, 150, 153, 155, 157, 170, 180, 184–186, 192
insider 87
outsider 32
quo 11, 38, 96, 126, 184, 192
tribal 121
steering committee 107
Stewart, Irene. *See also* Native Americans
strategies
conventional 172, 175, 180
unconventional 13, 171–172, 175, 180
Student Nonviolent Coordinating Committee 90
suffrage. *See also* elections and voting 19
support
emotional 82
survey research 42, 50
sustainable
economic development 47–48, 168
growth 57
use 144–149

Teamsters. *See also* union 102, 154
temperance 24, 38
theory
building 5, 38, 51
grounded 40–43, 50, 56
rational choice 152–154, 180
Third World Movement 91Tourists 116
toxic waste sites, unmonitored 61
tradition 41, 118, 122, 140, 147
traditional religious practices. *See also* American Indian Religious Freedom Act; Great Spirit; Medicine Men; Mother Earth; Native Americans 18, 38
treaties
Treaty of Guadalupe Hidalgo 17, 18, 38
tribal
council 22, 38, 45, 118, 121, 150
federal relations 35, 48–56
membership 121, 150
policies 189
sovereignty 33, 107–108, 110, 114, 121, 128–129, 150, 155, 161–162, 180, 185, 192
status 121

United States Government
Department of Commerce 5, 46, 72, 173, 75, 84
Department of Defense (DOD) 6, 38,
Department of Energy (DOE) 6, 7, 38, 76
Environmental Protection Agency (EPA) 62, 170
Forest Service 147, 150
House of Representatives 31, 38
General Accounting Office (GAO) 60, 61

union
  organizers 86
United Church of Christ Commission for Racial Justice 60
United Farm Workers 86, 90
United Nations
  Conference on Environment and Development 66
  Fourth World Conference on Women 188
uranium 6, 38, 64, 99, 167

violations
safety 166
volunteers 44, 107, 154
voting. *See also* elections and suffrage 19, 20, 22, 36, 38, 103, 122, 172

waste, solid 143
Waste Isolation Pilot Project (WIPP) 7, 38, 64
water
  pollution (*see also* Clean Water Act) 6, 38, 143quality standards (*see also* Clean Water Act) 34, 38
  rights
weapons research 7, 38
weaving 16, 38, 148
Wheeler-Howard Act. *See* Indian Reorganization Act (IRA) 20–21
wilderness areas 142, 144, 146
women
  domestic, role of 83
  gender roles 39
  of color 4, 9–10, 23, 27, 29, 37–38, 41, 50, 53, 56, 68–70, 77, 94, 100, 110–112, 117, 125, 128–130, 138, 140–141, 143, 145, 150, 157, 159, 180, 182–186, 189, 191–192
  political elites 40
  state legislators 31
working class 2, 6, 27, 38, 41, 63, 67, 87, 87, 100, 130, 140, 156, 184
workplace
  hazards 64
  safety laws 100

Young Democrats. *See also* party 86, 91

For Product Safety Concerns and Information please contact our EU
representative GPSR@taylorandfrancis.com
Taylor & Francis Verlag GmbH, Kaufingerstraße 24, 80331 München, Germany

www.ingramcontent.com/pod-product-compliance
Ingram Content Group UK Ltd.
Pitfield, Milton Keynes, MK11 3LW, UK
UKHW022117190926
475702UK00013B/1494

* 9 7 8 0 4 1 5 6 5 0 5 2 6 *